Contracting for Business Success

Andrew Cox and Ian Thompson

Thomas Telford

Published by Thomas Telford Publishing, Thomas Telford Services Ltd, 1 Heron Quay, London E14 4JD. Tel. 0171 665 2464; fax 0171 537 3631.
URL: http://www.t-telford.co.uk

First published 1998

The Contract Selection Toolkit by the same authors is available from:
Earlsgate Press, Boston, Lincolnshire, UK.
Internet: http://www.btinternet.com/~earlsgate.press

Also available from Thomas Telford Publishing:
Strategic procurement in construction. Andrew Cox and Mike Townsend, 1998.
ISBN 0 7277 2599 8

Distributors for Thomas Telford books are
USA: American Society of Civil Engineers, Publications Sales Department,
            345 East 47th Street, New York, NY 10017-2398.
Japan: Maruzen Co. Ltd, Book Department, 3–10 Nihonbashi 2-chome, Chuo-ku,
            Tokyo 103.
Australia: DA Books and Journals, 648 Whitehorse Road, Mitcham 3132, Victoria.

A catalogue record for this book is available from the British Library

ISBN: 0 7277 2600 5

Printed and bound in Great Britain by Redwood Books, Trowbridge, Wiltshire.

Contents

Preface

This book summarises a way of thinking about contractual relationships which companies involved in construction (especially those involved in the search for more effective sourcing and supply relationships) can use to achieve business success. It also has strong links with two companion volumes. The first is by Andrew Cox and Michael Townsend, entitled: *Strategic Procurement in Construction* and the second is by the current authors entitled: *The Contract Selection Toolkit*. This book needs to be read in conjunction with both these publications because, taken together, they constitute the heart of the thinking about better practice in construction procurement that has been developed at the Centre for Strategy and Procurement Management (CSPM) at the Birmingham Business School.

This work on construction procurement management, of which this volume is only a part, has been funded primarily by the following organisations: BAA, the Construction Round Table, Railtrack and London Underground. The funding for the research, that informs this volume, was generously provided by London Underground and Railtrack. We are extremely grateful to these two organisations for providing us with the finance to allow this work to be undertaken, although they are in no way responsible for the ideas and conclusions that have been arrived at here. The sins of omission and commission that have been made are due entirely to the authors' endeavours alone.

The authors would, however, like to thank unreservedly all those who have contributed directly or indirectly to the research by which this book has appeared.

Andrew Cox and Ian Thompson
Birmingham
December 1997.

Introduction

On Power and the Effective Management of Contractual Relations in Construction Procurement

---

Who is this book for?

For those who flick through its covers and peruse the contents page, it will be clear that it is a book that the major clients in the construction industry ought to find of interest. The reason for this is self-evident: it is because clients ultimately finance construction work and provide the financial incentives that 'energise' construction supply chains. Without the desire of major public and private sectors clients to award contracts to main contractors there would be no construction supply chain at all.

So *Contracting for Business Success* is clearly of interest to those end-users who commission construction work. It is, however, also one of the most important topics of interest to everyone upstream and downstream in the myriad of supply chains that constitute the construction industry. The reason is, again, self-evident. It is because most clients in construction do not undertake building as the primary way in which they make a living. As a result, they rely on others (professionals, main contractors and trades sub-contractors) to provide the knowledge and expertise to allow them to sustain the construction and maintenance of any of the physical assets and infrastructure that they require to allow them to make money.

Construction is, therefore, for most clients, at best, a complementary or, in the case of the majority of large or small construction clients, a residual activity in the overall scheme of things. Therefore most clients, whatever their total level of spend on construction, depend on contracts to ensure that those practitioners in the construction industry, which they rely upon, do actually deliver what is expected of them – on time, to agreed quality standards and within agreed costs. Obviously, clients retain some professional expertise in-house to help them with the management of their projects and expenditure but, unless the spend is a critical part of their business activity, for most clients the actual design and physical building, maintenance and repair of their infrastructure is undertaken by external providers.

Contracts with clients are, therefore, of primary importance to all of those actors within construction supply chains who make their living through the provision of construction-related services. It is of equal importance to the trade sub-contractor, or to the architectural or engineering professional, as it is to the main contractor, that they each negotiate and sign contracts that allow them to achieve business success. This book should, therefore, be of interest to everyone in the construction industry because contracts are the legally binding (*sic*) documents through which the mutual exchange relationships between the physical supply of goods and services and value take place in the construction industry.

Contracts touch and affect everyone in the industry and not just because they are the legal framework within which an exchange relationship, that passes a physical good or service between one actor in exchange for money (value) from another, occurs. There are two other primary reasons why contracts are important. The first is that they constitute one of the most contested areas of interaction within the industry. Many commentators have commented on the adversarial nature of contractual relations in the industry and have contended that mechanisms ought to be put into place to provide more equitable ways of contracting to enable the industry to be less claims based and more effective and efficient. This book takes issue with this way of thinking primarily because it fails, in our view, to focus on the primary importance of contracts in any business relationship in general and in construction in particular.

A second reason why it is important to focus on contracts in construction is because they constitute a *power relationship*. What this means is elaborated in more detail below but it is sufficient, at this stage, to highlight a key point which many commentators fail to understand when they discuss the endemic problems of contractual conflict in construction. This is the fact that the contract is the focus of conflict because it is the basis on which power is manifest in the multiple exchange relationships that exist within the construction industry.

It is our view that this aspect of contractual relationships has not been properly understood in many of the descriptive legal and professional texts that discuss construction contracting. As a result, this descriptive, as opposed to analytic, way of thinking leads to relatively naive views about what can and cannot be achieved within the industry through the re-writing of contractual forms, or through the discovery of alternative ways of managing disputes. In the pages that follow each of the traditional and current forms of contract that are commonly in use in the industry is analysed, especially from this analytical perspective of the power in the exchange relationship that the contract codifies. Having analysed the power inherent within the contractual form suggestions are then made for construction supply chain participants about which forms of contract may, or may not, be most suitable for the management of their particular business purposes.

The most common forms of dispute resolution and conflict avoidance used in the industry are also analysed. The critique provided of these techniques is also based on the relative capacity of the particular techniques to provide construction supply chain participants with the appropriate mechanisms to achieve their business goals. Overall, however, it is argued in this book that focusing on contractual solutions to the problems in the construction industry is likely to be a waste of effort. The reason for this, it is argued, is because to do so is to misunderstand fundamentally both the structure of power within the industry and how contracts can and should be used to achieve specific business goals.

In what follows in this introduction an analytical way of thinking about construction contracting for business success is briefly outlined. This outline is based on a summary of the essential elements of the arguments about power in construction that are developed in Cox and Townsend's *Strategic Procurement in Construction*. Following on from this, a second introductory section briefly outlines the major arguments contained in each of the five major Parts that make up this volume.

The basic argument presented in *Strategic Procurement in Construction* is that most of the current ways of thinking about how to improve efficiency and productivity in construction are misguided. This, it is argued, is because current government-backed reform proposals (such as Latham, Levene and *Setting New Standards*), as well as those emanating from the industry itself, are either self-serving or fail to grasp the nettle of the need to ensure that some interests in the industry will have to lose in order that others can be winners. It is contended, therefore, that there can be, despite what the Latham Report implies, no 'win-win' solution for everyone in the industry. This is because efficiency and productivity improvements in business have, through the ages, always involved a process by which some interests (either through a technical innovation or through their ability to monopolise market power) have been able to control the appropriation and accumulation of value from supply chains for themselves, against the interests of other actors in an industry.

If this conclusion is historically and scientifically correct, then it allows us to understand that most of the groups that are involved in fashioning reform agendas for the construction industry do so in a void that fails to address the issue of power.

By power in this context one means: *"the ability of an individual or an organisation to own (or to control) specific resources or assets (goods, service or know-how) within a particular construction supply chain, in such a way that it allows them either to appropriate the majority (or a substantial share) of the value that flows through the chain, or to determine the allocation of value to other participants throughout the entire supply chain."*

If one reflects on this way of thinking about the nature of relationships within the construction industry it becomes possible to see that much of the prevailing analysis of the industry starts from a fundamentally flawed premise. To take the Latham Report as an example it is clear that this Report was bound to fail in implementation, quite simply because it started from entirely the wrong intellectual foundation. The goal was to arrive at a solution that would encourage all of the actors in the industry to develop a collaborative (team-based) approach to supply management. The problem with this goal is, however, that the construction industry is not team-based, nor is it ever likely to be.

The reason for this is self-evidently obvious for those who have the eyes to see. It is because the construction industry is a seething mass of adversarial conflict, in which specific professional groups and interests, alongside white collar entrepreneurs representing main contractors and specific trades craftsmen and unskilled workers, vie to control and own the appropriation and accumulation of the value that flows through the industry. To expect all of these conflicting groups to work amicably together to pass value to clients, in the interest of the productivity and efficiency of the national economy, is clearly naive.

It is clear, as has been argued in *Strategic Procurement in Construction*, that there are nearly always structural reasons why it is possible for the long-term collaborative relationships that reduce contractual conflict (which the Latham Report argues for) to be created in any supply chain or industry. The major reason why long-term collaborative relationships occur is normally because of the creation of a governing relationship in a supply chain, based on the ability of one player in that supply chain to be able to control the allocation of value to the players within it. Normally, as has been the case in a number of Japanese industries, this governing relationship is based on the development of a clear ***coincidence of interest*** on the part of the players in the chain.

What generates the possibility of a coincidence of interest occurring in any supply chain? Is it, as the Latham Report argues, because of the ability to develop trust or is it, perhaps, because there are other more base reasons? It has been argued in that the primary reason why any coincidence of interest occurs is likely to be rather more base, than due to any altruistic attachment to partnerships of equality, openness, honesty and trust. Put quite simply, long-term collaborative relationships usually work because they are based on both sides to the agreement obtaining that which they value. It is, however, important to recognise that in this coincidence of interest it does not necessarily follow that both sides to the relationship will achieve benefits in equal measure. This is because of the relative degrees of power that exist within any business relationship in a supply chain.

Because there is normally an asymmetry of power in business relationships, all that is required is that the parties to any exchange relationship obtain more from accepting the deal than they believe would be the case if they did not participate. For those who are the suppliers of goods, services and know-how in any supply relationship the trade-off might be between a low income and a relatively permanent guarantee of employment. In other circumstances the trade-off might be a guarantee of regular profits and of continuous work that other players in the market cannot achieve, in return for open disclosure of one's working practices. Whichever of these it is (and there are many other forms of trade-off possible), it is clear that most people will only enter into long-term exchange relationships if there is a relatively high probability that to do so will lead them to obtain more than if they pursued a more opportunistic and arms length relationship with others and played the field.

What all of this implies is that people, by and large, are economically rational, wherever they come from. It is not that the Japanese are more economically rational than those in the West, which explains why they have longer-term business relationships in many industries. It is, rather, because the governing elites in that society have designed some of their industrial power structures in such a way as to make it possible for some actors to control and own the key resources that flow through supply chains, in such a way as to allow them to offer long-term relationships to others, with a high degree of certainty that what is offered can be delivered. This implies that, in the absence of this certainty in Japan, the long-term relationships that have been created may well begin to break down in the future. This is because Japanese people, ultimately, obey the same economic laws of rationality as the West. If there is no certainty of demand and if the industry is highly fragmented, then, it is likely that Japanese construction supply chain players will behave more opportunistically than they have done in the past.

As concluded in *Strategic Procurement in Construction*, and in Andrew Cox's *Business Success* (1997), a number of important conclusions flow from this insight that supply chain thinking is, ultimately, about the management of power in economic relationships. It is quite easy to see how long-term relationships could be created in the UK construction industry, but it is unlikely that it can be achieved voluntarily by the current players in the market. For collaboration to occur on a long-term basis the first thing that the government would need to do would be for it to select no more than five or six of the current main contractors and give them the bulk of the government's building work on a relatively permanent basis and for the major private sector companies with a high level of construction spend to do the same.

This action would immediately create an oligopoly structure in the UK industry and provide the regularity of demand that would allow these companies to begin to engineer quasi-vertical (or in some cases vertically integrated) supply chains. Companies with the levels of certainty of demand that this would give them would, then, be able to eradicate opportunism from their corporate supply chains almost over-night and either keep any consequential waste reduction and value improvement for themselves, or, perhaps, pass it on to their clients. Whichever of these two scenarios occurred would, of course, be dependent on the balance of power between public and private sector clients and the oligopolistic main supplier structure that this action created.

Clearly, this oligopolistic structure could be created by the concerted actions of major public and private sector clients, it is doubtful, however, that it could be achieved using the methodology recommended in the Latham Report. The reason why the Latham Report is incapable of delivering the structural transformation in the UK construction that is desired is, of course, self-evident. The reason is politics, ideology and power. The problem for the Latham Report is that it was commissioned by a government that had an almost religious devotion to neo-liberal ideas about economic efficiency and was entrusted to an intelligent politician who understood that, in the absence of compulsion, the only alternative is persuasion.

Unfortunately, despite the elegance of the ideas that the report contained, it is our view that the intellectual basis behind the report's thinking is fundamentally flawed. The reason for this view is the belief that persuasion is rarely the reason why people do things in business. People do things in business and in life, normally, because they must, not because they chose to do so. People can only do what they would like to do when their basic economic needs are, by and large, met. In an industry like construction, in which there are few real barriers to market entry and in which there is a surfeit of supply and inadequate demand encouraging people to be other-directed is a recipe for a myriad of implementation committees; it is not a recipe for fundamental and profound change in industry culture or structure, as has been witnessed through the post-Latham initiatives. The reason for this is not because of the malicious intent of people, but because the structural circumstances within which individuals find themselves forces them to behave in certain ways.

Many readers may have been affronted by what has been said above because they can point to a similar number of UK examples of clients, contractors and sub-contractors achieving considerable cost improvement and efficiency in quality and time through closer working relationships. Let us look at this in more detail. Why is it, despite what has been argued here, that some companies, notably BAA, McDonalds and Rover in the UK, have been able to achieve considerable success through the adoption of a more long-term collaborative approach to their construction spend? The problem is that anecdotal evidence from these industry players has suggested that significant performance improvements can be arrived at through long-term relationships (and/or through standardisation of building methods). While it may be true for these particular organisations, it does not follow that this constitutes grounds for arguing that everyone in the industry can, or should, operate in precisely the same way.

Having studied the activities of a number of companies

involved in significant performance improvement in construction in recent years, it is evident that there are clear and unambiguous structural reasons why these improvements have been possible. Furthermore, given this fact, there are also clear and unambiguous reasons why it is highly unlikely that these performance improvements can be achieved by the majority of players in the UK construction industry, whatever the Latham Report believes to the contrary.

The major case studies analysed in *Strategic Procurement in Construction* involved companies which have a regular and relatively high level of demand for construction work. Only one of the companies analysed was primarily a contractor, while all of the remainder were clients. Having analysed their recent actions, it is clear that the clients, through using a variety of fairly common-sense methodologies, have found a variety of ways to achieve significant improvements in time, cost and/or quality. So far so good, so why cannot everyone do the same? The reason is simple enough. The companies concerned have been able to engineer their improvements by providing an appropriate trade-off that creates a *coincidence of interest* with their construction suppliers.

This trade-off is simple enough to understand. The clients analysed have been able to guarantee a regular and high level of demand, in return for a willingness by preferred contractors (and/or sub-contractors) to be prepared to accept a degree of structural control and dominance by the client over their normal way of doing business. This is a relationship of pure power, in the sense that the supplier has to be prepared to forego or reduce the potential for opportunism against the client, in return for the buyer's promise of work in the future. In the current climate in the UK industry, with low operating margins and few technological resources that allow suppliers to monopolise the supply market against potential clients, it is hardly surprising that such regular spending clients have been able to bend some of the suppliers in the industry to their needs.

Two interesting questions flow from this. Is this approach appropriate for all clients to use, under all circumstances in construction? What does this case study evidence tell us about the historic construction procurement competence of major clients in the past? Taking the second of these questions first, it is clear that historically most construction clients were, and in some cases still are, relatively incompetent in understanding the first principles of business leverage and the effective management of supply chain power relationships. Everything that has been done in recent years, under the banner of 'best practice' in construction in the UK, is nothing more than the adoption of a common-sense approach by a number of clients. These clients, interestingly enough, do not appear to have arrived at this need for change through any first principled based intellectual efforts of their own, but have rather been driven towards a more logically rational and appropriate approach because of intensifying competitive pressures in their own industry sectors, or due to privatisation introducing market disciplines to formally ossified management structures and practices.

Rather than applauding most of these companies for developing world class practices, as many seek to do, our own view, therefore, is that they are doing nothing more than simply adopting a common-sense approach to effective supply chain leverage. This is because they each possess a regular process spend, that provides them with the potential to leverage their supply chains effectively. Now, for the first time, they have been able to recruit, or free, one or two able people, who understand what business is about, and who have begun the process of turning the potential power that they have always possessed into actual power. What is surprising is not that this is now being done, but that it has taken so long for these companies to wake up to the opportunities that have been so manifestly before them.

What is perhaps more worrying, however, is the still large number of private sector clients (and contractors) in the construction industry with this type of potential, but who do not appreciate the fact. These private sector companies do not understand the concept of appropriateness and still have some way to go in actualising the potential power that they possess. The situation in the public sector is even more lamentable. It is clear from our research that nobody in this country has even begun to address the issue of the potential buying and leverage power that resides in the public sector, whether in construction or elsewhere in the economy. The commitments to PFI, CCT and market testing that have bedevilled the public sector construction spend in recent years are, in our view, symptomatic of a complete blindness, by governments and civil servants alike, to the nature of buyer and supplier power in supply chains. It is clear that there is a significant ignorance in government to the possibilities for significant cost, quality and time improvement in public sector construction procurement through effective leverage, based on the horizontal, vertical and quasi-vertical integration of construction supply chains. The potential power that exists is not understood and nor are the mechanisms to actualise it.

Turning from why there is some evidence of better practice in construction procurement in the UK, to the question of whether or not everyone is capable of copying what the likes of BAA, Rover and McDonalds have been able to achieve using long-term collaborative relationships, it is clear that a pessimistic view must be taken of the general possibilities for the majority of actors in the construction industry. This will clearly be a disappointing conclusion to have arrived at for those, like the supporters of the Latham Report, who are wedded to simplistic stakeholding and partnership sourcing solutions to all business problems. It is, however, not necessarily a cause for concern that long-term collaborative relationships cannot be used by the majority of actors in the construction industry. The intellectual basis of this way of thinking is summarised below.

The major reason, why the practices adopted by Rover, McDonalds and BAA are not available to the majority of actors in the construction industry is because they do not occupy the same structural types of supply chains as these three companies and they do not have the same power potential within the industry as these exemplar cases do. This means that to pursue long-term contractual relationships is likely to be a highly inappropriate thing for many actors in the industry to do to achieve their own business goals.

Let us consider this for a moment in more detail. It was argued earlier that many of the better practice clients in UK construction operate with a high level of construction spend (relative to the volumes in the industry as a whole) and they have a fairly regular level of demand for similar types of construction activity within that spend. This provides them with a unique position within their own particular construction supply chains. It gives them a high degree of potential power over their suppliers within the exchange relationships that occur whenever they come to market. To put it at its simplest: the supplier needs the buyer more than the buyer needs any particular supplier.

This is clearly a very useful position to be in when conducting any business-related negotiations. This is because, if the buyer organises his negotiations and selection processes professionally he will be able to leverage the supplier most effectively. Obviously, the most effective way for the buyer to exert leverage over the supplier (i.e. bend the supplier to the buyer's wishes) is for the buyer to limit the number of suppliers who are awarded contracts (being careful at the same time not to create any dependency relationships), in such a way that a group of preferred suppliers are created. It helps, of course, if the preferred suppliers are desperate for the work and are not in a position to turn it down (i.e. they are dependent on the buyer) or in a position to behave opportunistically when other work comes along.

Clearly, in the structural supply chain position described above, the buyer is in a position of structural power *vis-à-vis* the suppliers in the industry. It would be surprising, in these circumstances, if the buyer would need to rely over much on the enforcement of performance improvements or compliance with performance benchmarks from the supplier through the threat of contractual terms and conditions. The buyer rarely needs any recourse to the contract (i.e. it can stay in the draw), because the power in the relationship is so much with the buyer that the supplier knows that even if they win a battle over a particular claim (legally speaking), they will risk the loss of the war (i.e. they will not be awarded any more work in the future).

But is this situation of relative power and dependency the condition under which the majority of building work in the UK operates? It is our view that this condition, based on clients possessing a structured and regular process-spend for similar types of construction products and services, is unlikely to constitute more than 25% of the total construction market in the UK. This is because the majority of construction expenditure is for one-off project-specific items that are often bespoke to the needs of individual clients and which have to be built under very different ground conditions and contingent supply chain circumstances.

In such project-specific supply chains everyone, including the client, has an objective incentive to behave opportunistically. The reason for this is because the buyer rarely returns to the market more than once for the product or service and, consequentially, lacks either the expertise or the competence to understand in detail what it is that is being purchased. As a result the buyer, having no regularity of spend to offer suppliers, lacks the power to be able to enforce his wishes on the supply chain.

Similarly, because the majority of clients find themselves in this position and have to rely on the supply industry for their competence, in this type of supply chain the suppliers are potentially more powerful than the buyer at the point at which any contractual negotiation takes place. This is, however, both a strength and a weakness for suppliers. This is because in this type of supply chain structure, it is normal that the supply industry becomes highly fragmented, with many different actors (engineers, architects, project managers, main contractors etc.) seeking to claim that they have the supply chain management competence that the client needs. There is, as a consequence, likely to be a surfeit of supply, both of professional and managerial advice, as well (because of the low financial and technical barriers to entry) as a surfeit of trades people to undertake the work.

The result of this fragmentation and intense competition is that no one group of suppliers is able to win sufficient amount of the business to allow them to effectively control and manage their own supply chains effectively. There are two reasons why. In the absence of an ability to dominate the supply industry no one is able to so control demand to allow them to rationalise and leverage the supply industry to provide a performance improvement barrier to entry. Second, because buyers recognise their own relative incompetence and impotence, they have a vested interest in encouraging a highly competitive and fragmented supply-base to exist. This is because, economic theory tells them that where there is a multiplicity of inter-changeable supply, margins will be keen and they will be able to behave opportunistically in the use of contracts that force the supply industry to take and to manage all of the risks inherent in any construction project.

The suppliers, given that there are so many of them, have no real choice other than to acquiesce to this type of supply chain structure. They have to accept the effective leverage position that they find themselves in and try to pass any consequential risks that flow from it down the supply chain to their own sub-contractors. This is because, in the absence of regular demand, they must behave as opportunistically against their own suppliers, as the clients are behaving towards them. It is hardly surprising, therefore, in this type of supply chain that legal claims are rife and contractual disputes are endemic to supply relationships.

Our own view of this structural reality is somewhat different to that of some recent commentators. Rather than being concerned with this situation we accept it is a fact of life. In an absence of a willingness by government to induce, or major industry players to voluntarily agree to, the creation of an oligopolistic industry structure, there seems little prospect of goodwill destroying the structural properties which underpin the project-specific construction supply chains that dominate the industry.

Obviously, while there is tremendous scope for clients and their suppliers in process-based supply chains to engineer efficiency improvements collaboratively over the long-term, it is wholly unrealistic to assume that this will be equally as possible in the project-specific supply chains that constitute the bulk of current construction activity. Thus, rather than writing another book that hopes to achieve something that we believe is unachievable and inappropriate for most of the actors in the industry, this book is devoted to explaining how participants in the construction industry might usefully think about how they can adopt *a contracting approach to business success*. This contractual approach to effective leverage and supply chain management in construction is, we believe, the one that is the most appropriate way for all of those who, because they operate in a project-specific supply chain structure, are unable to develop the more long-term collaborative approach to effective leverage and supply chain management.

Given what has been argued above the structure of this

companion volume is easily understood. The book seeks to explain, for the majority of those in the industry (who are not in the fortunate position to be able to use collaboration as the most effective means of achieving leverage of their business goals), what is the most appropriate way in which to think about the use of contracts in the search for business success in construction procurement. This is achieved in five major sections:

Part A considers current contracting practices in the industry and what the role of the contract is in construction supply chains. It also.reviews the current types of contracting methods that exist by the way of an introduction to subsequent Parts.

Part B analyses in detail the current contractual forms used in the industry and assesses the balance of power within these forms. It makes recommendations about which may, or may not, be the most appropriate to use to achieve specific valued outcomes for particular practitioners in the industry under specifc circumstances.

Part C applies the same type of analysis to the public sector and explains why the government may have so significantly missed its way in thinking about effective leverage, through the use of such recent practices as PFI, the EU Procurement Directives, CCT and market testing/outsourcing.

Part D looks at the current tools and techniques in dispute resolution and makes recommendations about the most appropriate way to use these techniques. It considers the whole area of conflict management and it questions the validity of dispute avoidance measures.

Part E summarises the major arguments in the book and indicates which types of contractual forms may be the most appropriate for actors in construction to use under specific circumstances. It also provides a brief introduction to how one might begin to think about the use of optimal forms of contract for specific problems.

Part A

The Current Situation

Chapter 1

Current Contracting Practice in UK Construction

Introduction

The construction industry is in a mess. Few are able to dispute this fact: it is not just that the industry's productivity and service are so relatively poor compared with other industries, nor the fact that the industry continues to maintain a long-developed and tarnished reputation for failing to deliver client satisfaction. The UK construction industry is packed with in-fighting, self-interest and inefficiency. This is not just the case among the trades disciplines, it is so throughout all tiers of the industry and its professional institutions.

Recent opportunities to address some of these issues have failed to make any impact on the industry and its service delivery. These opportunities include major public reviews of government and industry[1] as well as private self-assessed analyses by institutions and their membership[2]. Despite the burgeoning literature reviewing the industry's supply-capabilities, there are few texts that share the same conclusions or the same remedies. Each suggests it has the answers and disregards what the others have to say.

Moreover there are few signs of any change. Despite the multitude of reviews and the employment opportunities this has generated for consultants and academics, the industry does not appear to want to take note. Those who do make sympathetic sounds end up being condemned by their rivals for 'marketing hype': cynicism and apathy prevail.

Meanwhile, clients still fail to receive satisfactory service and value for money, while contractors and consultants continue to struggle to survive commercially. The fight for control of ever-scarce resources continues apace: margins are virtually non-existent and loss-leaders are common, while claims and disputes proliferate. The construction industry is no longer a sound investment; it is too highly geared, too unprofitable and too absorbed in the fight against insolvency.

So what is wrong with the industry?

Surprisingly there are few disagreements about the symptoms of the industry's problems. Some of the more common issues are illustrated in Figure 1.1. The list is not exhaustive but it does offer a flavour of the state of the industry. The real debate, however, concerns the nature of the 'cure'. How should the industry be re-structured to deliver better service to its customers while ensuring the key players retain healthy and profitable margins?

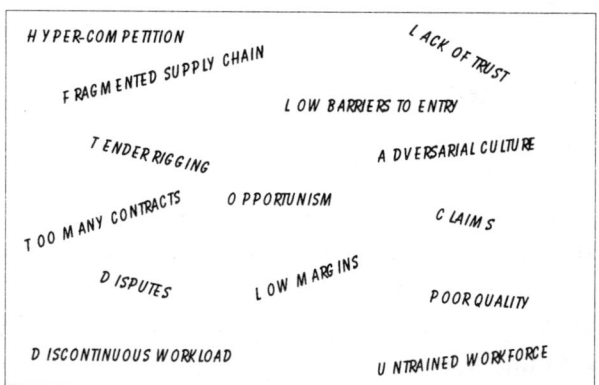

Figure 1.1: Some Problems in the Construction Industry.

The issue of 'what to do' with the industry remains contested. There have been numerous recommendations from many sources, but few have been implemented successfully and few have delivered the necessary changes. Perhaps unsurprisingly, each of the proposed sets of recommendations differ; it seems that there are few who can agree on an appropriate course of action to improve the industry. It is not the intention of this book to examine these proposals in detail. There are, however, a number of common criticisms that can be levied on the approaches:

- most reviews have attempted to concentrate on the industry's symptoms rather than address the root causes of the problems, with the result that most proposals are cosmetic rather than effective;
- most analyses have concentrated on the 'what is now' scenario without considering how the industry could be changed or re-structured;
- those which are promulgated by specific institutions and/or trades confederations are biased towards the support of their own membership and do not necessarily consider the needs of the industry or its client-base; and
- those reviews which cross all divisions of the industry have broadly attempted to offer 'something for everyone' proposals without addressing the fundamental issues at the heart of the problems[3].

About this book

While it is important to understand the current state of the industry and the context in which this book is written, this book does *not* seek to address all of the fundamental problems of the construction industry. Instead it specifically considers the contractual issues of procuring construction works. However, to understand fully these issues and why certain conclusions have been drawn, it is imperative to understand both the structure of the industry and the underlying root causes of current problems. Ideally, this book should be read in conjunction with *Strategic Procurement in Construction* by Cox and Townsend[4], which considers some of the broader issues of market structure and supply-capability throughout the supply chains which make up the construction industry.

This book complements that study by considering the appropriateness of an approach to value-appropriation based on contracting and the commercial aspects of business transactions in construction. It is primarily focused on the interface between the commissioning client and the industry's supply base. Unlike other construction management texts, this is considered from a business perspective, rather than focusing on the technical or legal content of the transactions.

The book is structured in five major parts. The first introduces the reader to current contracting and procurement practices. It has two main objectives: firstly to bring the reader up to speed with contemporary concepts in contracts and procurement; and secondly to challenge the existing 'dominant paradigms' of current practice in the industry. Rather than describing the 'what is now' scenario, we draw on experience and insights gained from outside the industry to review the objectives of contracting within construction.

The middle three parts (B–D) proceed to describe existing practices by looking at the most common contracting methods and some of their popular forms of contract, as well as procedures for resolving disputes and conflict. These parts consider these provisions in sufficient detail to allow the reader to be familiar with each of the mechanisms, but without labouring on the minutiae of detail that could be expected in a legal commentary or exposition. The final part of this book (Part E) summarises the previous parts examining the *appropriateness* of each mechanism as a commercial instrument and to consider under which contingent circumstances they would be most applicable.

The primary purpose is to determine how contracts and the relations between contracting parties can be re-engineered to deliver the business objectives in a better way. Thus, it is the authors' intention to determine what it means to be *Contracting for Business Success* in the construction industry.

The State of the Industry

It has been established that one of the most telling indicators of the UK's economy is the count of tower cranes on the City of London's skyline taken from the roof of St Paul's Cathedral. Although the accuracy of this statistic should not be taken too seriously, it is indicative of the 'feast or famine' extremes that ravage the industry and their proximity to the 'boom or bust' cycles of the national economy[5].

This uncertainty and instability feeds all aspects of the industry. In times of high demand, the supply capability is found to be sadly lacking: labour prices rocket and materials shortages prevail. It is not uncommon to have labour-only suppliers move from site to site on weekly bases looking for increasingly exorbitant pay rises, or for clients to find themselves embroiled in 'Dutch auctions' for key materials such as brick, concrete and steel, where the highest bidding site receives the stockists' supplies, irrespective of who placed the order. The fight for control of scarce resources causes prices to spiral upwards and quality levels to drop. At the same time, there is an influx of unskilled and inexperienced suppliers establishing themselves to meet the high demand. The consequential effect is that unwary clients find themselves procuring construction works from an incompetent supply-base and trying to depend on contractual conditions which are dated and inappropriate.

The problem is that this is not a static state: the industry moves in and out of these cycles. In times of low demand, suppliers are forced to compete for scarce and discontinuous workloads. Clients are keen to buy at the lowest price and suppliers find margins eroded. In order to reduce their on-costs, suppliers shed staff and resources through redundancy and outsourcing. Contracts are won on inaccurate tender estimates and/or loss-leaders, thus placing extreme pressure on site staff to recover revenue through change orders, claims and even disputes. Meanwhile the supply capability diminishes as suppliers become more and more dependent on a network of subcontractors to whom they may pass unfair risk burdens and slow payments. Under this type of market there is a gradual shake-out of suppliers as insolvencies increase and the principles of 'natural succession' prevail.

Thus neither extreme provides a competent supply market from which clients can procure their construction works effectively. In both extremes the self-seeking interest of short-term 'opportunism' prevails – and so it should for these market conditions! Where are the incentives for suppliers to conduct business in any other way? Long-term policies and strategies cannot be pursued while the fear of insolvency remains in each quarter of trading.

The statistics to support these descriptions speak for themselves[6]. The last construction boom in the UK was in 1990 when annual output reached *circa* £56 billion. At the time of writing in 1997, output is only £49.8 billion which, if accounting for inflation, is nearly 20% below 1990 levels. Since this boom, approximately 500,000 construction jobs have been lost[7] and there have been more than 16,000 company insolvencies. Current tender prices are only just beginning to resemble those offered in 1989; meanwhile contractual claims and disputes continue to proliferate. In Chapter 12, it is demonstrated that at least 10% of all business in construction is being contested at any given time. Furthermore the supply market remains highly fragmented: only 5% of construction firms (by number) have more than seven employees and yet they conduct nearly 70% of the national workload (by value).

Current Contracting Practice

Given this level of fragmentation and the view expressed in the introduction that this is inevitable for a substantial proportion of the industry, the importance of *competent contracting* practice becomes self-evident. Construction operations are no longer integrated in the same organisation; the majority of operations are conducted beyond the boundary of the client's firm with contracted suppliers; business operations become the product of a commercial transaction between two (or more) contracting parties. A successful supplier in the construction industry needs to ensure it has a high level of *contracting competence* to survive. Equally, any client who wishes to procure works from the industry and who is unable to develop the collaborative-leverage approach outlined in the introduction, also needs to have a correspondingly high level of *contracting competence*.

What is 'contracting competence'?

For those readers who are only used to thinking that contracting concerns a decision between ICE or JCT forms of contract, the notion of *contracting competence* may appear to be somewhat academic and far-fetched. Far from this! Contracting competence involves all the day-to-day operational decisions that practitioners need to make when faced with the practical circumstances of their business transactions. It is based on an understanding of *appropriateness* for the contingent circumstances faced by practitioners. This is best considered in terms of the client, as illustrated in Figure 1.2 overleaf.

Contracting competence is derived from the appropriate application of contracting mechanisms to deliver the business objectives operationally, within specific circumstances. These circumstances vary according to the nature of the transaction and the content of the construction works, as well as the structure of the market and the supply chain characteristics within it. In other words, they are *contingent*. Of the total universe of possible contracting applications, only some or one of these applications will be appropriate to deliver the client's business objectives in these contingent circumstances. Contracting competence is about understanding when different applications can be most appropriately applied.

For those practitioners interested in understanding how to link types of contracts to specific contingent circumstances in the construction industry, reference to our companion guide on contract selection is recommended[8]. This operational guide comprises a rigorous methodology for practitioners to select and implement the most appropriate form of contract for the specific circumstances of their commercial transaction(s). It provides guidance on each of the industry's standard forms of contract for both building and civil engineering work, as well as the use of amendments and bespoking.

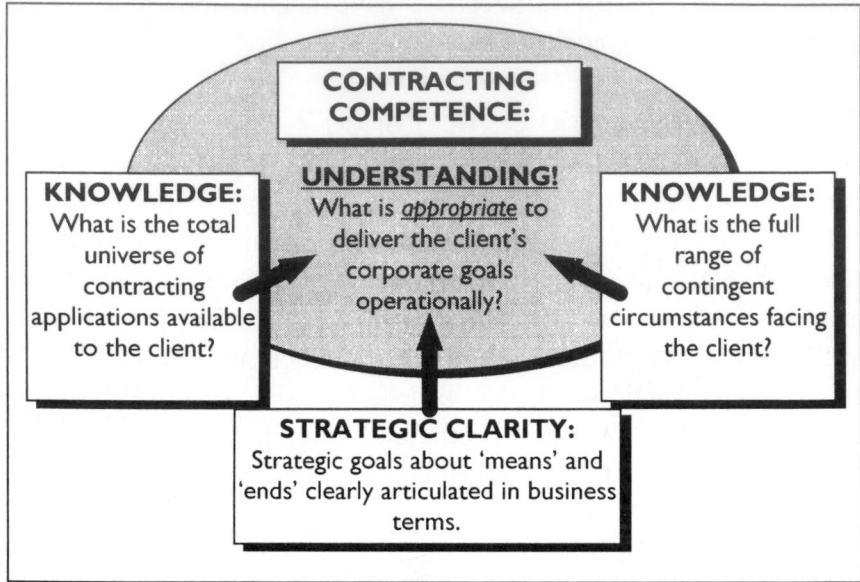

Figure 1.2: Understanding Contracting Competence.

Adapted from: Cox[9], passim.

Chapter 3 examines the construction supply chain in greater detail; however full competence can only be gained by knowing all the permutations of the market and understanding their effects on the commercial aspects of business transactions. In Parts B–D of this book, we begin to examine the universe of contractual applications and the circumstances in which they are appropriate or not. Pragmatism suggests that not all permutations can possibly be listed or described. What the practitioner requires is *understanding* about when the circumstances dictate an application is appropriate or not. This cannot necessarily be taught, although readers are referred to Cox's work on business success[10] for further understanding of the concept.

What contractual applications exist?

Traditionally there has been too strong a focus on the specific conditions of contract that apply between contracting parties. A confusion between 'means' and 'end' has developed; it must be recognised that the contract and the particulars of the relationship between the parties only serve as the means to the end of the transaction. This point is picked up in the next chapter which considers the purpose of contracts and their role in commercial transactions.

This over-emphasis on the contract has had a detrimental effect on the delivery of construction services to customers. This, in turn, has effected the behaviour and performance of suppliers which ultimately impacts on the performance objectives of the overall works. In chapter 2, we advocate that the industry should re-focus on performance and use optimum contracting arrangements to meet these end-goals.

In the manifold industry reviews there have been a considerable number of suggestions to improve contractual and procurement arrangements. These have included recommendations for new contracts, new legislation and new ways of relating and behaving. Subsequently there have been several conflicting suggestions concerning the efficacy of these recommendations and their application to the industry. While different groups are promoting their own 'new' forms of contract[11], others are suggesting that *relational* practices, with an emphasis on collaboration and 'partnering'[12], are the solution to the industry's ailments. There is no singly agreed 'best practice' and the industry is lacking advice as to which contracting methods are most appropriate for whom and on what occasion.

Meanwhile the industry continues with its traditional behaviour and approach to procurement. Construction projects are procured on a one-off basis with little regard to future needs or supply development. Thus the supply-base is bespoke to the needs of each project: the parties come together for the duration of the project and then go their separate ways. There is little benefit gained from shared learning or synergy between parties as fragmentation has led to individualism and self-seeking interest. Relationships are confined to the discrete duration of the contract; trust, commitment, reciprocity and other behavioural aspects associated with long-term collaborative relations remain absent. The parties rely on the formality of the governing contractual conditions: the focus of the transaction is *contractual* rather than *relational*.

Form of Contract:	Number of variants:	Current status:
JCT'80	6 variants + partial design supplement	15 amendments in use.
JCT'81	1	9 amendments in use.
JCT IFC'84	1	9 amendments in use.
JCT MW'80	1	9 amendments in use.
JCT MC/87	1	2 amendments in use.
FIDIC, 4th Edition	2 construction contracts	In use.
ICE 6th Edition	1	1 set of corrigenda in use.
ICE Design-Construct	1	In use but under review.
ICE Minor Works, 2nd Edition	1	In use.
NEC, 2nd Edition	6 options	3rd Edition under review.
GC/Works/1, 2nd Edition (Minor Works)	1	In use.
GC/Works/1, 3rd Edition	3 variants	In use.
DEFCON 2000	2 construction contracts	In use.
IChemE Red Book, 3rd Edition	1	In use.
IChemE Green Book, 2nd Ed	1	In use.
PSA/1	1	In use.
CCS	1	Little used.
Total Number:	31	

Figure 1.3: Table of existing forms of contracts[13].

To add to the industry's problems, there are at least 16 recognised standard forms of contract in regular use at present, as well as many other less popular conditions (see Figure 1.3). There is little or no guidance available on which type of contract is the most appropriate under which conditions. The guidance which does exist[14] relies heavily on the traditional approaches to procurement using some of the existing forms of contract but with no regard to the nature of demand and the existing supply market conditions, or the effect of a governing relationship between the parties. The result is a lack of clarity and general confusion concerning which contracting strategies are appropriate at any particular instance. This has the consequential effect of driving clients along a narrow path that suggests the form of contract alone is the basis of the transaction.

For example, at the time of its privatisation in 1991, the Central Electricity Generating Board (CEGB) had approximately 100 approved construction contracts for use in its industry. These included contracts for professional services, mechanical and electrical contracts and specifically drafted bespoke forms of contract. Many of these contracts were redundant, some were never used and some were unknown.

The problem was that not everyone used the same conditions in similar circumstances. The result was that different procurement practices were carried out in different parts of the organisation with no real guidance on which form of contract should be used and when. Suppliers were not always familiar with the contract they were being asked to price, which led to inefficiencies while another learning curve was climbed. Furthermore, suppliers supplying the same goods or services to different parts of the organisation were being asked to do so under different contractual conditions, which could often be managed in very different ways. These wide-scale discrepancies simply added unnecessary costs to an already costly process.

Following privatisation, National Power plc has attempted to rationalise this list from the 35 regularly used contracts down to a workable number for the majority of their transactions[15]. Scottish Hydro-Electric plc has also reduced its number of ex-CEGB contracts with an increasingly heavy emphasis on the NEC (see chapter 8).

The power generating companies are not the only clients to have a wide range of contracts. Britain's custodian of the railway infrastructure, Railtrack plc, has an approved list of contract forms to be used for construction works. The list comprises at least 20 different standard forms of contract and a further 15 specially drafted bespoke contracts[16].

The issue for large and diverse companies such as these, is not how the number of contracts can be reduced to make contracting practices simpler to manage, nor is it how each of the circumstances requiring different contractual applications can be covered by a different contract. The issue is whether the companies can develop an optimum number of contractual applications to deliver the corporate objectives and to know under which circumstances they are most appropriate to use.

Failure to understand this will fail to instil contracting competence which, in turn, will fail to deliver the corporate objectives of the business.

At present there is general confusion throughout the industry at the expense of customer frustration and dissatisfaction. Meanwhile the cost of disputes and contractual claims continue to increase.

Summary

It is clear that the industry's existing contracting practices are sub-optimal; they are the seed-bed of conflict and dispute, rather than being an optimum mechanism for the discharge of efficient and effective construction works. There is a need for clients to develop *optimum* contracting strategies which will successfully serve business needs and ensure 'best practice' construction procurement for the client. In recognising the need for sustainable markets, cost reductions and efficient working practices, appropriate contracting strategies will recognise the needs of the supply base too.

The remainder of this book sets out a framework in which the reader may consider the strategic intent of the client's business, the range of contingent circumstances facing the client and the universe of contracting applications that exist. It is hoped that this book also imparts understanding about the *appropriateness* of the applications in order to raise general levels of contracting competence in the industry.

Chapter Notes

1. Refer to: Latham M. (1994) *Constructing the Team: Final Report of the Government/Industry Review of Procurement and Contractual Arrangements in the UK Construction Industry* HMSO, London; Levene, P.

(1995) *Construction Procurement by Government: An Efficiency Unit scrutiny,* HMSO, London; HM Government (1995) *Setting New Standards: A strategy for Government Procurement,* HMSO, London, Chapter 4; and Audit Commission (1996) *Just Capital...Local Authority Management of Capital Projects,* HMSO, London.

2. Refer to: The Institution of Civil Engineers (1996) *Whither Civil Engineering?* Thomas Telford, London; and The Royal Academy of Engineering (1996) *A Statement on the Construction Industry,* London.

3. For further discussion on the efficacy of the Latham proposals refer to: Cox A. & M. Townsend (1997) 'Latham as half-way house: a relational competence approach to better practice in construction procurement' *Engineering Construction and Architectural Management,* Vol. 4, Issue 2, pp. 143 - 158.

4. Cox A. & M. Townsend (1998) *Strategic Procurement in Construction,* Thomas Telford, London.

5. For a good description of the impact of boom and bust cycles on the construction firm refer to: Hillebrandt P. M., J. Cannon & P. Lansley (1995) *The Construction Company in and out of Recession,* Macmillan Press, London.

6. All these statistics are taken from the DOE's *The State of the Construction Industry Report,* Issue 7, February 1997, unless otherwise indicated.

7. *Management Today,* February 1996.

8. Thompson I. & A. Cox (1998) *The Contract Selection Toolkit,* Earlsgate Press, Boston, UK.

9. Cox A. (1997) *Business Success: a way of thinking about strategy, critical supply chain assets and operational best practice,* Earlsgate Press, Boston.

10. *Ibid.*

11. For example: Abrahamson M. (1995) *Risk, Management, Procurement and CCS* in J. Uff & A. M. Odams (eds) *Risk, Management and Procurement* Kings College London; and Barnes M. (1996) 'The New Engineering Contract An Update' *The International Construction Law Review,* Vol. 13, Part 1, pp. 89 - 96.

12. For example: Baden-Hellard R. (1995) *Project Partnering: Principle and Practice* Thomas Telford, London; and Bennett J. & S. Hayes (1995) *Trusting the Team: the best practice guide to partnering in construction* University of Reading.

13. Thompson I. & L. Anderson (1997) 'Optimal Contracting Strategies in Construction' 6th Annual International IPSERA Conference, Naples, Italy, (25 March).

14. Clamp H. & S. Cox (1989) *Which Contract? Choosing the Appropriate Building Contract,* RIBA Publications, London.

15. Institution of Civil Engineers' Factsheet prepared by N. Shaw and G. Warren (1995).

16. Thompson I. & L. Anderson (1997) *Op. cit.*

Chapter 2

The Role of the Contract

Introduction

This chapter provides a framework around which the rest of this book is built. For some construction professionals this will be spelled out in material that they are already familiar with as part of their professional training. Nevertheless the authors believe that, despite its familiarity to some, there are important concepts which need to be emphasised and reiterated at the outset of this book. Some, but not all, of the material presented in this chapter is accepted practice and it is important for those immersed in the management of day to day construction activities to understand that other professional disciplines approach supply chain contracting issues in different ways.

Based on a synthesis of the current better practices, this chapter returns to first principles to offer what is considered to be a rigorous review of the role of the contract in commercial transactions in construction. In places there is reference to the legal capacity of contracts; like the rest of this book, this chapter is not offered as a legal treatise, where the reader requires more detailed information on issues of construction law, it is recommended that reference is made to more suitable texts which are focused on this[1].

There are two principal considerations regarding contracting that are of key interest to business professionals:

1. the purpose(s) of a contract;
2. the role the contract should play in the business transaction.

Since contracts require *consideration* (i.e. an exchange of some material value), by nature they are a central theme to all business transactions, whether perceived or not. Contracts provide the legal framework for all buying and selling of goods, services, products, projects, processes, rights, corporate acquisitions, etc. In the construction industry there is an over-emphasis on the conditions of contract and their effect on business activities. The following text is written with construction in mind, but could equally be applied to other supply chains.

What is a Contract?

To fully understand the nature of contracts and the issues associated with their administration, our understanding needs to be broader than the sterile definitions offered by text books. Lawyers often describe contracts in terms of their principal constituents (tenets), but again this is insufficient for those wishing to apply good practice and achieve their corporate business goals.

There are several types of contract and many ways of contracting, the key is to gain sufficient understanding to be able to recognise what is *appropriate* for the particulars of a business transaction. This theme is central to this book and will be repeatedly considered throughout this and subsequent chapters.

In the following section the nature of contracts is considered:

*   by their definition;
*   by their constituents;
*   by looking at their types (and their alternatives); and
*   by their function (i.e. what they do).

Understanding contracts by definition

Few text books offer a formal definition of a contract. One definition may be considered as: *"a legal agreement to exchange value between two (or more) parties"*; this incorporates all the fundamental requisites of a contract. The principal issue surrounds the *value exchange*, which is another way of describing a business transaction. In this way contracts can be seen to be at the heart of all commerce. It should be noted that usually contracts are formed between two parties, although there are exceptions in the case of tri-partite or consortia agreements, *inter alia.*

Contracting comprises the administration and management of contracts. This could be considered to include all aspects of their development and fulfilment, including:

- **drafting the contract:** this involves writing the terms and conditions in an unambiguous and legal manner. It can be done *unilaterally,* by one party only, *bilaterally* through negotiation between the two parties to the intended contract, or *multilaterally* through industry representatives or committees. Since the contract will become a legally-binding document requiring the parties to comply with its express requirements, both parties will be keen to ensure that the contract terms are written in their favour and this can be the source of early conflict in a new business relationship;

- **formation of the contract:** this occurs at the point of agreement – i.e. when an offer has been accepted;

- **performance of the contract:** this means conducting the business for which the contract was agreed – e.g. the supply of 1000 tonnes of raw materials, or the construction of a chemical process plant, etc.;

- **discharge of the contract:** i.e. completion point – when the contract has been fully discharged, all parties have fulfilled their responsibilities and the contractual conditions no longer apply; e.g. the 1000 tonnes of raw material have been supplied *and* payment received for them, or when the chemical plant has been constructed *and* full payment has been received. Note: in both these examples there has been a *full exchange* of some material value.

From the above list it is clear that there is an administrative process involved in contracting which is over and above the required actions of the contract. This process has costs attached to it which are referred to as *transaction costs* (see Figure 2.1). Other transactions costs might include the identification, accreditation and selection of suppliers and/or performance monitoring and feedback. Obviously if the contracted activities were carried out in-house, rather than being contracted to the external supply market, these transaction costs would not necessarily apply. Many outsourcing exercises fail to consider the full costs of contracting-out. These full costs are referred to as the *total cost of ownership* and, as well as the transaction costs, include the whole-life costs of operation, maintenance, decommissioning and close-out.

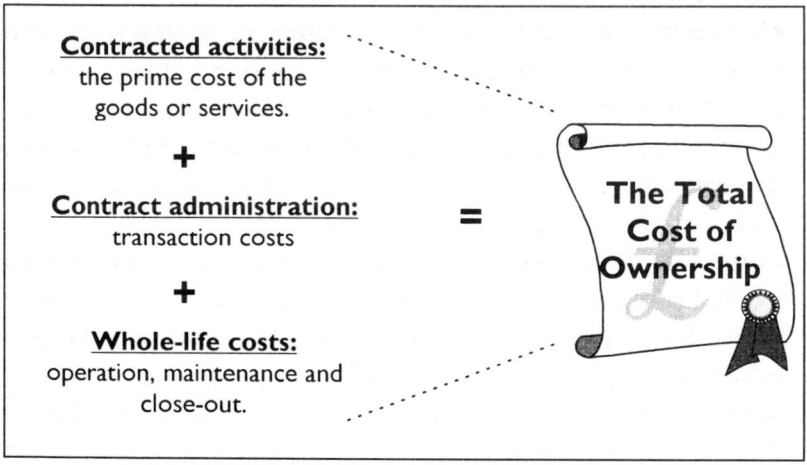

Figure 2.1: The Costs of Contracting.

Understanding contracts by their constituents

To determine whether an agreement or a legal document is a contract, English law requires the existence of certain pre-requisites. Without these pre-requisites there is no contract and the parties' actions are not covered by the law of contract. The following bullet points provide a very simple view of the principal tenets of contract:

- **Offer:** an offer is tendered by one party to the other as a pre-contract proposal. It has all the requisites of a contract but for the agreement between the parties. Offers can be accepted or rejected by the recipient while they apply.

- **Acceptance:** in order to reach an *agreement*, the offer must be matched with a corresponding acceptance. The terms of the acceptance must perfectly reflect the offer. Problems can occur if the offer has lapsed or been 'revoked' (withdrawn) before acceptance, hence nullifying any potential agreement. Similarly, if the acceptance does not match the original offer, it is said to be a 'counter-offer' which, in turn, requires acceptance or rejection by the other party. In pre-contract negotiations there can often be several offers and counter-offers before the parties agree the terms of business.

- **Consideration:** refers to the exchange of value. It does not have to be a fair amount, or even for money; it is part of the pre-contract discussions to decide the terms.

- **Intention to create legal relations:** the courts tend to view all business transactions as though the parties concerned wished to have created legal relations. The only situations where this is not applicable is in domestic arrangements (i.e. between husband and wife) or for 'a gentleman's agreement' which is *binding in honour only*. This is an important consideration when considering the merits of *relational contracting* (i.e. business carried out on the basis of a governing relationship where no written contract exists, as discussed in chapter 10). The key consideration revolves around whether it is the intention of the parties to 'contract' and form legal relations, or whether this is specifically being avoided in the interests of promoting close-working relations.

  As an aside, it should be noted that it is possible for subsidiary companies to have legal contracts with other subsidiaries of the same parent company and, furthermore, it is not unknown for them to pursue disputed contractual matters in the courts.

- **Other legal requirements:** these include *legality* (the contracted requirements must be legal), *possibility* (the requirements must be physically possible to perform) and

capacity (the parties to the contract must have legal capacity -
i.e. be in their right mind and of the 'consented' adult age).

Understanding contracts by their type

Very few commercial contracts are totally identical; nevertheless
the same principal tenets must be found in them (offer, acceptance,
consideration, intention, etc.). Contracts will be distinguished from
each other by their terms and conditions. Effectively these are the
'rules, regulations and procedures' governing the business
transaction. They may cover issues of responsibility (i.e. who does
what), timescales, risk apportionment (in the case of intervening
circumstances beyond the control of the parties), payment terms,
quality issues and/or rights of redress (if one party fails to do what
it contracted to do).

The terms and conditions can be written down within the
contract documentation, be referred to in a supplementary
document, or be present by implication (in the case of a governing
statutory law). Written terms and conditions are referred to as
express terms, whereas those present by implication are referred to
as *implied terms*.

An *oral contract* will cover the terms that were agreed verbally
between the parties and be backed up by any governing statutory
laws. Although oral contracts save a lot of time and money (in
terms of saved bureaucracy) there are many potential risks,
especially if the parties have interpreted their responsibilities
differently. This can lead to further problems if a dispute arises
over these differences, as it is difficult to support a contested issue
when there is little or no evidence of the contract. Consequently
most commercial contracts are expressed in writing in one form or
another.

Written contracts may include:

- **Bespoke Contracts:** these are unique one-off 'specials'. They
  have usually been drafted by lawyers and/or procurement
  professionals for the specific circumstances of a commercial
  exchange. They are not usually based on any known or
  established set of terms and conditions; i.e. they are drafted
  'from fresh', which has both advantages and disadvantages.

The main advantage is that the contract can be specifically tailored to the needs of the business transaction, i.e. it *can* be 'fit-for-purpose'. However there are many additional costs associated with this method: it costs a lot to employ lawyers to draft the contract, the contract is rarely suitable for other transactions and the other contracting party is unlikely to be familiar with the terms (hence a learning curve exists). These disadvantages mean that there are considerable *transaction costs* associated with bespoke contracts; unless the contract is going to be of considerable value (sufficient to justify this expense) it is unlikely that bespoke contracts will be optimum for the transaction.

- **Standard Forms of Contract:** a standard form of contract is a published set of terms and conditions which is well known and established within an industry. For example, the Chartered Institute of Purchasing and Supply publish several standard forms for all sorts of goods and services (e.g. IT services, consultancy, repair and maintenance works, etc.). The advantages are obvious: the terms and conditions are known by all and thus the transaction costs are greatly reduced. The possible disadvantage with them is that they are 'generic', where specific clauses may be more preferable.

  It should also be remembered that the standard forms of contract in themselves are *not* contracts! Despite the fact they have been multi-laterally agreed by industry committees, they still need to form part of an offer and acceptance by the specific parties to the contract. Parts B and C of this book examine common contracting methods and the industry standard forms of contract in greater detail (e.g. the ICE and the JCT forms, etc.).

- **Purchase Orders & Invoices:** these are common for commodity 'spot' purchases. The purchaser simply fills in the order form and the supplier supplies the goods with an invoice. Most forms have standard terms and conditions printed on the back. When the purchaser sends an order it is just an offer and, in return, the supplier may send back an invoice with different terms and conditions on the back of it. Technically this constitutes a counter-offer. For example, the different forms may have differing payment terms on it (a purchaser may demand 90 day payment terms, a supplier may hope for advance payment!). What follows is called the 'battle of the forms', as counter-offer followed by counter-offer is mailed between the parties' offices. The danger is that when the business transaction is actually performed, it is unclear which terms and conditions apply!

- **Flow charts:** this is a rare and radical means of expressing the terms and conditions of a contract. The New Engineering Contract (NEC) was designed using flow charts (and still refers to them as a guidance document, see Chapter 8), while Max Abrahamson has recently developed the CCS contract which is based wholly on flow charts[2]. The idea is that instead of prescribing what the parties should or should not be doing, they are guided through a 'process' by following the flow chart. One problem with this is the lack of clarity that occurs when the actions of one party (or both) diverge away from a position in the flow chart. The use of this system is still extremely rare and probably unlikely to become very popular.

A Supply Chain View of Contracting

Having established a better understanding of what a contract is, it is necessary to begin to consider contracting in the context of the whole supply chain (see Figure 2.2). In this illustration there is a 'chain' of activities and/or processes which transform raw materials into finished goods/products/services for use by the consumer. The chain is usually very complex and comprises a 'network' of activities and processes (i.e. it is rarely just one simple linear chain). Similarly it does not have to govern the passage of physical materials; it could be intellectual property, as in the case of consultancy services or research. This is discussed in greater detail in Chapter 3 which looks at the construction supply chain as the combined flow of materials and services to clients on a discontinuous project basis.

Usually there is more than one organisation involved and hence there is a need for contracts to govern the commercial transaction.

In the case that one organisation owns the full process, the supply chain is said to be *vertically integrated*. (As a slight aside, there is the notion that within organisations internal contracts can be formed [the *nexus of treaties* view] but these will not necessarily be legal contracts if there is no intention for the parties to form legal relations).

The supply chain focuses on the flow of material to the end-consumer. It is matched with a value chain flowing in the opposite direction, i.e. the flow of money from the end-consumer. (Note: this is not to be confused with Porter's value chain which is an internal view of organisations[3]).

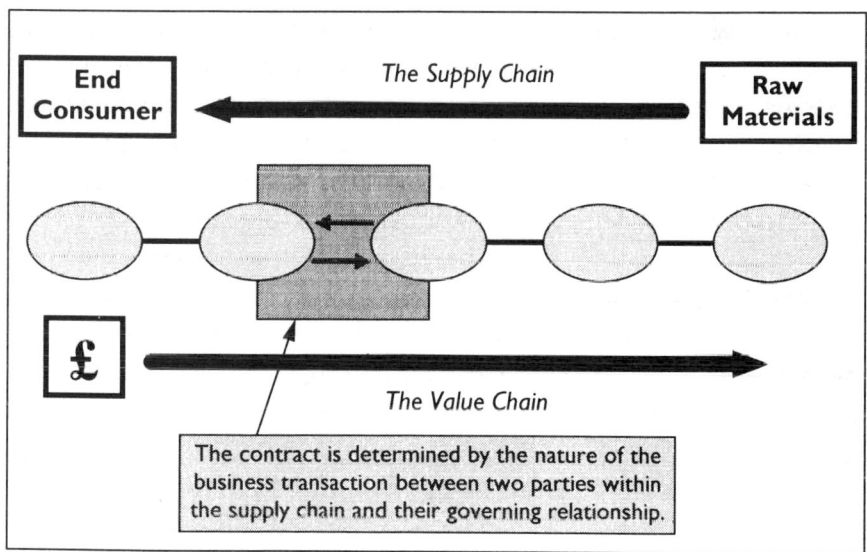

Figure 2.2: A Supply Chain View of the Contract.
Source: adapted from Cox[4].

For the purposes of this chapter and the remainder of this book, the 'contract' considers only one discrete exchange within the supply and value chains (as shown in Figure 2.2); it does not consider the full scope of the supply chain activities. This individual exchange between two parties is governed by a 'relationship' (which considers the behavioural aspects associated with power, trust, conflict, opportunism, etc. - and is usually more evident when there is on-going business between the parties in the longer term) and the legal contract itself.

This supply chain view helps us to consider the role of the contract in its full context. The key question is, given that there are different contracts and ways of contracting, what is the best way in which these can be arranged in order to achieve corporate goals?

What are the corporate goals?

As an important aside, a generalisation can be made concerning the differing views of the legal and business professions. The lawyer is intent on ensuring the contract is formed in the legally-correct manner and is usually focused on this as an end-state in itself. Conversely, the business professional is less focused on the exact nature of the contract and is more concerned about the business transaction and how it can fulfil the business goals. Within these differences of view, a distinction must be drawn between means and end: *the contract is only a means to an end!*

Given this, it is imperative to consider the business goals of the organisation, its strategic intent and, consequently, the business drivers that exist. Once set in this context, the discrete construction project can be seen as a contributor to business operations and not set apart from them (as has often been the case traditionally). This context is considered as a core of this book throughout the following chapters and, moreover, as the determinant of the terms and conditions under which construction is contracted.

What Functions do Contracts Perform?

A contract is a form of quasi-ownership. Although it may not necessarily form a strictly legal title of ownership, the contract can be used to own, control and manage transactions, processes and/or goods and services.

Business professionals are primarily concerned that the documentation of the commercial transaction achieves the business objectives that are required of it. As already stated this can cover many aspects and, not least of all, the contract provides a *record of the agreement*. As a documented transaction, the express terms assign risks and responsibilities to the parties to the agreement (i.e. it tells them who is responsible for doing what, as well as who is responsible *if* certain events occur). The contract also provides *legal cover* in the event of misunderstanding, default or error occurring (i.e. it protects the parties' interests). This may also be extended to cover the *rights of redress* which will detail what remedies there are if something goes wrong or if one party does not carry out its responsibilities - these rights might be in the form of pecuniary damages and/or penalties for default. A more positive measure, which is of increasing use in contracts, is the use of *incentives* to improve performance (as discussed in Chapter 10). Damages and penalties can act as disincentives which distance the parties. Incentives can (when properly applied) induce parties to collaborative working which, over a longer time horizon, can produce better results under certain circumstances.

One of the traditional views of a contract has been that it is a 'guarantee of compliance'. The view being among practitioners (but not the legal profession) that once the contract has been agreed, the other party must and shall comply with everything that is expressly required. In practice this rarely occurs, people are opportunistic and will cut corners to gain commercial advantages to suit themselves; in this respect the contract does not necessarily guarantee anything. Although technically this is not correct (as the contract can be reinforced by law), anyone who has had to rely on the force of law to ensure contractual compliance knows this is not an easy or enviable task.

Another similarly-traditional and misguided view concerns the behaviour of the parties. The contract has been viewed as a way of controlling opportunism and preventing 'self-seeking attitudes'; once the contract is agreed the other party is prevented from taking such actions. For the same reasons as the above paragraph, this view is questionable; the contract is not an efficient means of controlling parties' behaviour.

One slightly less common function of contracts is that of a *project management tool*. Contracts are regularly used to procure projects such as construction, IT and/or consultancy services, but they are rarely used as the mechanism which prescribes the parties actions in a programmed manner. However some standard forms of contract are meant to be used as a project management *aide memoire* (such as the NEC). They are based on the same principles as a flow chart and guide the parties as to 'what to do next'. This means that they are used and constantly referred to by the parties instead of being stuffed away in a drawer, only to be brought out and dusted down when problems occur. The arguments for this type of contract sound convincing (they encourage the parties to work together and thus reduce the likelihood of disputes) but are yet to be fully acknowledged. One common criticism is that they 'restrict' efficient and/or innovative working and cost the parties added overheads in following the prescribed guidelines in a narrowly bureaucratic fashion.

What are the Alternatives to Contract?

As has already been discussed, contracting is an indirect means of ownership. The most obvious alternative is direct ownership itself and this is called *vertical integration*. If an organisation produces a certain product then it is said to 'own' the process of production; its alternative is to *outsource* that process and contract with an external supplier to supply the produce instead. The decision whether to 'make' or to 'buy' is a fundamentally critical and strategic question confronting the organisation. This is discussed in greater detail in Part E of this book and also in Cox[4].

The advantages of owning a process or activity which produces goods and services are manifold: the organisation has direct control of the process (it can do what it wants), but this also means it has full responsibility for the process and therefore has to accept the risks which that brings. This also means that potential commercially-sensitive information and knowledge can be controlled and kept within the organisation and its employees. One further advantage is the reduced costs of transaction: the organisation does not need to spend time and money negotiating a contract or checking to ensure its compliance. Obviously there are disadvantages too, such as the costs of production and all its associated overheads.

It follows, therefore, that one of the critical factors when deciding to make or buy is identifying the *total cost of ownership* of a process (i.e. how does the cost of producing the goods in-house, with all the additional overheads that can carry, compare with the costs of contracting for those goods from an external supplier with the added transaction costs that incurs?), as was described in Figure 2.1 earlier. The fundamental key in the decision to make or buy, is to try to achieve all the benefits of vertical integration without the costs of ownership.

Ideal or Optimal?

Determining the most appropriate source for goods and services is, in many ways, only half the answer to best practice procurement management. Having decided whether it is best to own the process in-house (such as a direct-labour works service) or to contract from the external market, the next step is to consider the current sourcing arrangements and the changes that are required for implementation. This brings into consideration the practicalities of the market place and any limited resources or capabilities that exist. Whereas the 'ideal' practice is most desirable, compromise to account for practical restraints is required in order to achieve the 'optimal' (see Figure 2.3).

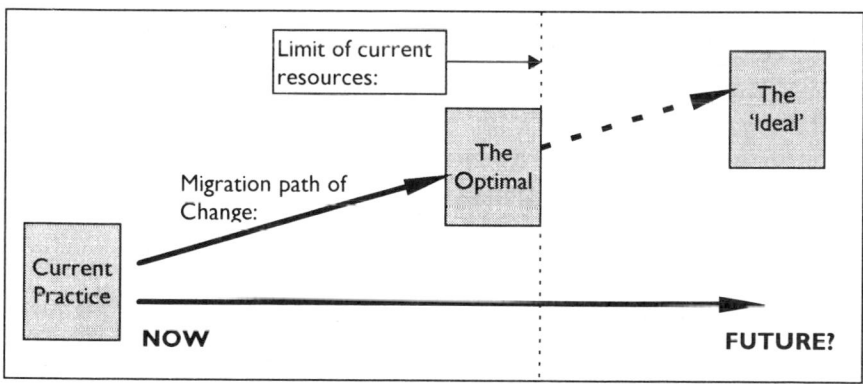

Figure 2.3: Achieving the Optimal Contracting Practice.

Adapted from: Cox[4]

The remaining parts of this book concentrate on the nature of the construction industry as it is today and how contracting practices can be improved to bring them closer to the 'ideal'. To achieve this with any realistic hope of success, it should be clear to the reader that, first, one needs to establish what is the 'ideal' and what is the gap that exists between it and current practice. This should then be followed by a change programme which understands the full context of the current practical circumstances and can aim towards achieving the ideal in order to reach the optimal state.

An example of this would be where an organisation is currently contracting with a large number of suppliers to construct an on-going programme of capital works using an antiquated form of contract. After assessment, it may be established that the 'ideal' situation would be to have a single dedicated contractor to undertake all aspects of the client's work, without the need for a written contract. However practical considerations might indicate that opportunism exists in the supply chain and that there is a risk of dependency growing with one supplier and that, furthermore, bespoke contracts would incur too many transaction costs. Thus the optimal in this situation may be to reduce the number of suppliers (e.g. to a 'parallel sourcing' arrangement) and to use a more appropriate standard form of contract for the transactions.

The Effect of Supply-relations

A final consideration for this chapter is the influence that the relational aspects of a business transaction may have on a contract and the effect to which it is discharged. These aspects have already been mentioned and include issues of:

- trust;
- opportunism;
- commitment;
- reciprocity;
- adversarialism;
- control;

- hierarchy;
- conflict;
- goodwill;
- dependency;
- power; and
- collaboration.

This list is not fully inclusive, nor will all the issues apply in any single instance of a commercial transaction. Nevertheless, each of these aspects will have an effect on the practical details of the transaction. For example, in his recent review of the industry, Sir Michael Latham was keen to establish fairness and mutual trust in all construction contracts[5]. While this was full of good intention, it can be argued that it failed to consider other relational aspects which prevail in the current industry, such as opportunism and adversarialism. Simply adding a clause in the contract to act in a spirit of mutual trust and co-operation will not over-ride the industry's dominant relational paradigm of opportunism (as is discussed in Chapter 8).

Had Latham addressed the structural properties of the supply chain, in order to reduce the fragmentation and individualism that prevail, his report may have been able to change the relational qualities and, in turn, influence the way in which contracting practices are conducted. In this context, the business-specific qualities of the exchange relationship being entered into have a determining effect on the contractual matters of the transaction.

These exchange relationship qualities of contracting are best synthesised in the term: *contractual relations*. It is the contractual relations between parties in a supply chain which determine how contracted matters are performed. This is discussed in greater detail in Part E of this book, however, Figure 2.4 illustrates the principal components overleaf. These comprise: the effect of a determining *relationship* between buyer and supplier; the effect of the prescribed *responsibilities* in the contract; the effect of the *risk* apportionment between the parties and the effect of the *reimbursement* mechanism for actions within the transaction.

The majority of literature in construction management has concentrated on the contractual aspects of risk, responsibility and reimbursement; whereas purchasing and supply texts have emphasised the importance of the supply relationship. Neither approach is sufficient or appropriate in its own right.

One of the main principles of this book is the need to emphasise how 'fit-for-purpose' contractual relations can be established so that the construction works are contracted to a competent supplier through an appropriate form of contract governed by an appropriate supply relationship. This is the principle of *relational competence*[6]. It, as mentioned before, is built on the foundation of *appropriateness* in order to develop what is optimal for the practitioner to implement and to achieve his/her corporate objectives.

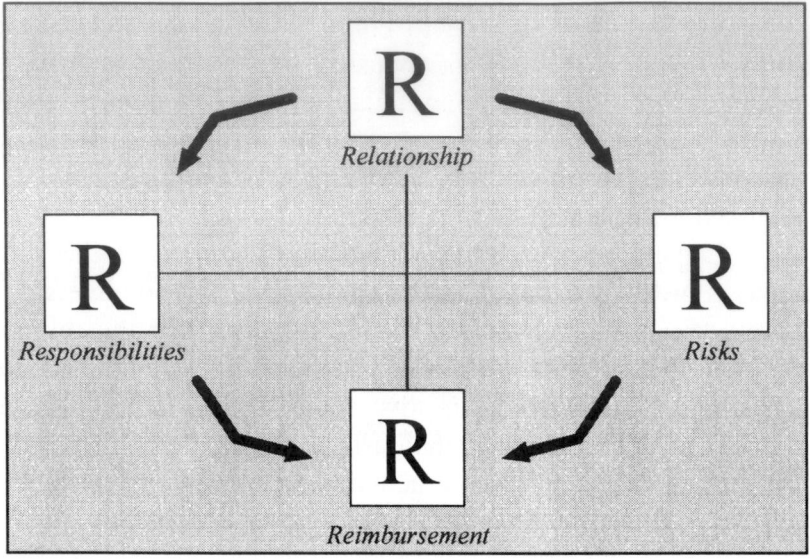

Figure 2.4: The 4 R's of Contractual Relations.

Summary

This chapter has spelled out the importance of contractual relations to externally sourced supplies, be they construction projects, goods or services. It has considered the need for a rigorous make/buy review to determine the 'ideal' sourcing strategy and the need to analyse the gap between it and current practice. Contracts can be expressed in a variety of ways, not just through the industry's standard forms of contract. The key is to develop contracting methods which are 'fit-for-purpose' and optimal to the delivery of the corporate objectives.

The remaining chapters in this Part consider the construction supply chain, its properties and characteristics (Chapter 3) and the popular contracting methods which currently exist in the industry (Chapter 4).

Chapter Notes

1. See for example: Uff J. (1996) *Construction Law, Sixth Edition*, Sweet & Maxwell, London.

2. Abrahamson M. (1995) 'Risk, Procurement, Management and CCS' in Uff J. & O. M. Adams (eds.) *Risk Procurement and Management*, Kings College, London.

3. Porter M. E. (1985) *Competitive Advantage: Creating and Sustaining Superior Performance*, Free Press, New York.

4. Cox A. (1997) *Business Success: A Way of Thinking About Strategy, Critical Supply Chain Assets and Operational Best Practice*, Earlsgate Press, Boston, UK.

5. Latham M. (1994) *Constructing the Team: final report of the Government/Industry review of procurement and contractual arrangements in the UK construction industry*, HMSO, London.

6. The theory of *relational competence* was developed in: Cox A. (1996) 'Relational Competence and Strategic Procurement Management: towards an entrepreneurial theory of the firm' *European Journal of Purchasing & Supply Management*, Vol. 2, No. 1 (March), pp. 57 -75.

Chapter 3

Supply Chain Contracting in Construction

Introduction

Chapter 2 introduced the concept of the supply chain and the role of the contract within it. This chapter considers the construction supply chain(s) in greater detail, in order to present a conceptual framework for the reader to consider the contribution and effect of *contractual relations* throughout the rest of this book. In industries other than construction, much discussion has been given to the concept of Supply Chain Management (SCM) and its contribution to business. Benefits have been heralded as including the eradication of waste and inefficiency, greater productivity, higher degrees of customer satisfaction and overall improved supply co-ordination, thus resulting in a 'leaner' more competitive business operation. The evidence of the success of this approach has come from manufacturing process industries, such as the automotive industry, where the adoption of Japanese 'lean supply' concepts throughout the supply chain is reported to have improved the competitiveness of all.

This chapter considers the merits of some of these concepts in an industry such as construction. Construction has many different attributes to general manufacturing and it is contended that the adoption of the same techniques, in totally different contingent circumstances, will not necessarily apply.

The chapter outlines the principal tenets of SCM and then proceeds to examine the nature of the construction industry. The authors argue that there are several supply chains combined and interwoven within the industry and that each has its own particular properties. The chapter proceeds to examine the nature of these properties from the issues of demand and supply and conclude that few of the generic lean supply concepts have merit.

Finally, for the construction practitioner who is unfamiliar with some of the terminology and concepts presented in this chapter, the authors have provided a 'further reading' section at the end.

What is Supply Chain Management?

There is a pressing need to clarify exactly what is meant by the concepts of the *supply chain* and the *value chain*. This is not just so in construction, considerable evidence can be collected from other industries and academic writings to demonstrate that there is a paucity of understanding about the terms. It is therefore essential to any text on the subject to define its terminology clearly.

Harland[1] has identified four specific descriptions of the supply chain. These are illustrated in Figure 3.1, however it will be clear to the reader that her taxonomy is, in essence, describing the same process (i.e. that of *supply*), albeit defined by alternative structures. The authors of this book contend that it is inappropriate to define the supply chain by the existing structure of the supply-base in terms of the number of players and/or the boundaries of their operations, *inter alia*, as this is rarely fixed in today's commercial environment of mergers, acquisitions and outsourcing.

Thus, for the purposes of this book, we have defined the supply chain to be the process of *activities* which transforms raw materials to finished goods and services for use by the end-consumer, irrespective of corporate boundaries (see Figure 3.2). A firm is situated within the supply chain and will contribute to the activities of the supply chain in order to make a profitable return for itself. In essence, it participates in the supply chain by buying goods and selling them on down the chain. This needs to be done in a competent manner in order to generate income (i.e. buy cheap and sell dear). The company may also contribute to the transformation process within the supply chain in order to earn greater returns, but this does not always have to be the case (as in the example of a materials stockist, *inter alia*).

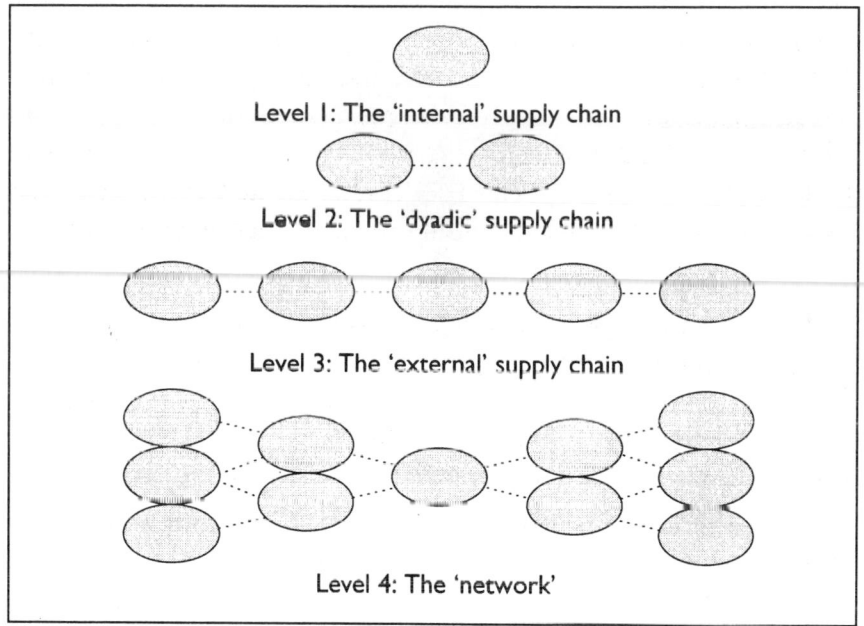

Figure 3.1: A Taxonomy of Supply Chains.

Source: Harland[1]

The value chain refers to the process by which money is exchanged through the supply chain in response to an initial supply offering. The exchange forms part of the contractual agreement identified in Figure 2.2 of Chapter 2. It flows in the opposite direction of the supply chain. Although it will share the same corporate structures as the supply chain, it differs in the resources allocated to each actor within the chains by the degree to which margins are appropriated and controlled. It therefore follows that successful firms are those which are located in such a position along the supply chain that they can control and accumulate maximum value for themselves (as demonstrated in Figure 3.2). This understanding is fundamental to this text, as it predicates all understanding of the power and value of contractual agreements.

Thus, in the following examination of standard forms of contract (in Parts B and C), the ability to appropriate value for the contract parties must be considered as an imperative to successful business.

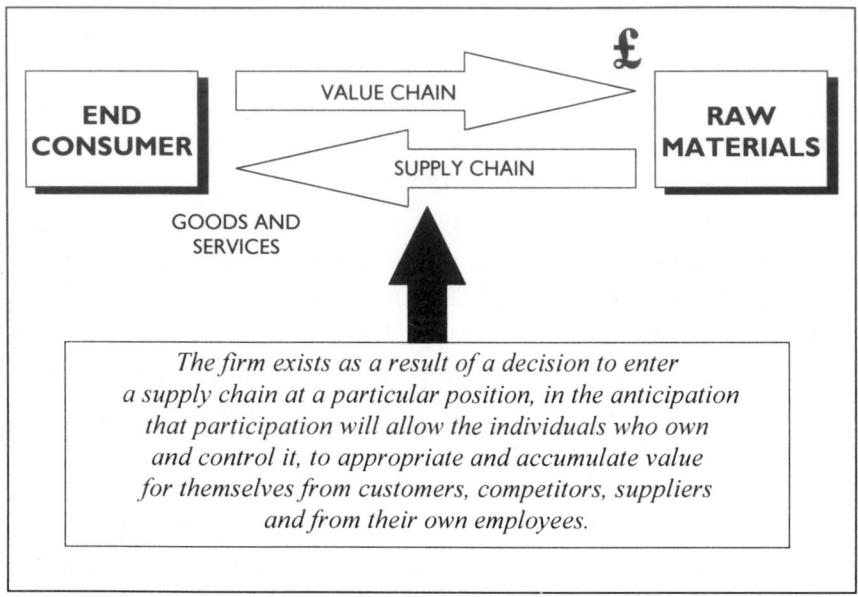

Figure 3.2: The Supply and Value Chains.

Source: Cox[2]

SCM has developed from the observation of Japanese management practices at large. Perhaps the most famous of examples is that of Toyota Motors and its tiered suppliers (refer to Case Study 1) which was reported in *The Machine That Changed The World* by Womack, Jones and Roos[3]. The persuasion of the argument is that because it appears to have worked for others, it should also work for us.

The benefits of an SCM approach

The following benefits have been reported from empirical studies of the adoption of SCM approaches in other industries. As these are listed, the reader is invited to consider their value and likely adaptation to the project-based construction industry:

- **Elimination of waste:** Through better management of the whole supply chain and the processes involved in it, wasted effort can be reduced. The classic example is the effect of *demand amplification* (known as the 'Forrester effect')[4] as irregularities in demand cause excessive over-ordering and unnecessary storage and distribution problems in the supply-base. This causes an unnecessary waste of resources which exacerbates further up the supply chain, away from the end-consumer. It is often fuelled by poor communication from party to party along the supply chain. Similarly uncertainty and conflict causes an additional strain on resources, as does duplication of effort and re-working of activities. The Total Quality Management (TQM) drive for 'getting it right first time' has been born out of this need to reduce wasted resources. While these are all desirable and common sensical practices, there is a limit on their effectiveness if the peaks and troughs of clients' demand cannot be smoothed.

- **Greater efficiency:** Linked to the above statements is the drive for greater utilisation of assets and resources, as well as the desire to abandon non-value adding activities. This drive has seen the growth of the Just-In-Time (*kanban*) concept. It also embraces the *core competence* thinking[5] which ultimately suggests that firms should only focus on what they are good at doing. While this concept has gained a great deal of popularity throughout Europe and the US, it assumes the firm's position in the supply chain is fixed and thus it fails to take account of the fact that any re-positioning exercises might yield better returns for the company.

- **Shared resources and capabilities:** As parties within the supply chain develop their shared interests and work together collaboratively, benefits can be gained from sharing resources and, over a longer time horizon, building up shared capabilities to supply goods and services to a common market. As resources are shared, the costs and overheads of those resources can be reduced. As the shared capabilities develop, supply chains can begin to compete against other supply chains supplying rival goods or services to the same market (e.g.

Nissan's supply chain competes against Ford's). Of course, this only works where suppliers are locked-in and unlikely to switch from working with one major client to working with another.

- **Synergy**: The sharing of resources and capabilities is based on close-working relationships between buying and supplying firms, often developed through joint development or problem solving teams. Information and shared learning exchanges are pre-requisites and, through the synergy that develops, new products and/or work processes can be innovated. On occasions, the degree of information exchange can go as far as cost transparency, although usually it will only concern projected workloads and future market developments.

- **Cost-base reduction:** This is the main attraction of all procurement and SCM activities. Supplier rationalisation programmes and 'tiering' have been the main vehicles for the reduction of transaction costs (as there are fewer interfaces and supply relationships to manage). Again, they have been based on the Japanese automotive model.

- **Customer focus:** Through a stronger focus on the needs of the customer which is communicated throughout the supply chain, there is a stronger focus on achieving the end-product in a manner that 'delights' the customer. As the above efficiency gains and cost reductions are realised, the benefits are passed down the supply chain to the end customer. As a result, the customer gains better value for money and the supply chain retains market share and volume of throughput (the emphasis switches from push or pull to one of *flow*). This has been particularly beneficial in high-volume turnover markets such as the automotive industry.

- **Greater competitiveness:** As a result of the stronger focus on the end-user and the added-value passed to the customer, suppliers who have adopted these supply chain management approaches have experienced a greater degree of competitiveness in their respective market places.

Case Study 1: Toyota[6]

The Toyota Motor Company was established on 28 August 1937 by Kiichiro Toyoda. The family-run business concentrated on producing military trucks for the Japanese Government, but achieved only limited success before suffering a collapse in sales in 1949. In the first 13 years, it made only 2685 cars [total] compared with the 7000/day produced at Ford's Rouge plant. The collapse in sales, mainly caused by political intervention throughout Japan, threatened to lay-off a third of its workforce and months of industrial disruption followed.

A settlement with the unions was finally reached in which flexible working practices and jobs-for-life were guaranteed. Following this, Kiichiro's nephew (Eiji Toyoda) and another production engineer (Taiichi Ohno) were sent to Detroit for three months to learn new production techniques. On their return, they concluded that simply copying Ford's system of mass production through vertical integration would not work back home in Nagoya. They began to re-consider the contingent circumstances facing Toyota and innovate their own manufacturing techniques; which later became known as the "Toyota Production System".

There followed a revolutionary change in operating practices. Ohno introduced elementary quality controls based on go/no-go switches thus reducing the amount of labour input required on each machine tool. A card system was established for parts inventory which acted as a rudimentary information system and a precursor to Just-In-Time delivery. Ohno also re-arranged the production line to manufacturing 'cells' based on a horseshoe pattern. By simply adding or subtracting workers to each cell, productivity was made to be flexible to the fluctuations of the market.

Toyota began to contract out an increasing amount of its operations from approximately 75% of its value-adding assembly operations conducted in-house in 1937 to approximately 25% by the late 1950s. Each outsourced supplier was 'tiered' and controlled through equity-sharing and parallel sourced operations, thus maximising the benefits of collaboration and competition. The structure enabled each tier in the supply chain to concentrate on a few suppliers at any given time and, thus, gain better control of the process and its outputs.

As this production system progressed through the 1960s, 1970s and 1980s, Toyota emerged as an increasingly competitive global player. It was during this period that it concentrated on quality management, continuous improvement, streamlined logistics, just-in-time delivery and improved production methods – all the hallmarks of today's 'lean thinking'.

By the late 1980s, Toyota was producing 4 million vehicles p.a. with 37,000 employees, compared to General Motors' 8 million vehicles with 850,000 employees. Despite only 30% of the added-value activities being conducted by suppliers, GM required 6000 purchasing staff, compared with Toyota's 337 purchasing staff responsible for over 70% of their added-value activities.

In the 1980s, Toyota began to diversify its operations out of the automotive industry and into household appliances, real estate, cotton and synthetics, as well as house building. Toyota first began making prefabricated housing in 1975 to demonstrate the applicability of the Toyota Production System in other industries. In 1987 it opened its Kasugai works, the first of three plants producing pre-fabricated houses. Since then, the housing business has grown steadily and it now produces approximately 3000 luxury house units *per annum* for the domestic market. The homes are built inside factories in very similar ways to Toyota's other manufacturing products. They are promoted and sold through an extensive network of dealerships in Japan and, once purchased, transported to site for final erection and assembly, where they are then fitted and furnished. Because of the way in which Toyota has applied a 'process' approach to construction, it has been able to gain all the benefits of the 'lean' supply chain in much the similar way as it achieves through car production.

But Does it Work in Construction?

The aforementioned list of the benefits of SCM and the example of Toyota in Case Study 1 are highly attractive and carry persuasive attributes. Everyone in the industry would wish to eliminate their own waste and inefficiency, reduce their cost-base and enhance their competitiveness by learning from their customers and/or suppliers. Indeed, this 'end' is a noble ideal, but it is questionable whether it is realistic. What if the elimination of your clients' wasted over-orders results in a direct reduction of your core business to them? Furthermore, is the added-value which is passed through to the customer at the expense of your own would-be profit? If either of the answers to these questions were affirmative, then the viability of some of these practices could be seriously undermined.

Even if the SCM concept works in other industries, more specifically, does it apply to construction? To answer this, we need to consider the properties of the construction industry by examining the nature of demand and supply.

Nature of demand

- **Project Specific:** Construction is often based on projects rather than on-going processes or continuous workloads. Something has to be built or renovated and the activities that comprise this are, in general, *discrete*; i.e. they have a distinct start and a distinct end. This means that the focus of the client's demand and consequently the supplier's contracted activities are project-oriented. Although some construction projects can be large and complex, compared with general manufacturing processes, the focus of activities is relatively short-term. This is further exacerbated by the team-based approach to construction, where each project requires the assembly of different team-players on a temporary basis.

- *Ad hoc Demand:* Linked to the above, clients' demands for construction services vary according to the nature of their primary business activities. Individual clients might not return to the market place for a considerable period of time. The combined effect on the domestic economy is one of irregular fluctuation; the demand is *ad hoc*, as discussed in the introduction to this book and the parallel publication by Cox and Townsend[7].

- *One-off Basis:* Each construction project has unique properties that, from a technical point, make the undertaking a 'one-off' experience. This is often an argument put forward in favour of individual project tendering, irrespective of the commercial gains evident from demand consolidation or the high level of transaction costs incurred by not consolidating demand. This argument also prevails against benchmarking industry standards as a like-for-like basis of comparison cannot be established. Again these factors are exacerbated by the team-basis of the industry. Different teams are required and invited to tender for different projects; thus there is little transfer of learning from project to project.

- *Technical Basis:* Construction is a technical industry which is dominated and led by engineers, architects and building technicians. Often the focus is on the technical merits of the construction project rather than the commercial attributes. Furthermore, few clients have internal construction expertise. From the authors' research to date, it is estimated that approximately 5% of UK clients (by number) actually possess internal capabilities (which equates to approximately 40% of clients by value of workload). This means that many clients are dependent on the professional advice of the supply-base. Few clients are able to direct and develop the supply-base to meet its needs and wants.

- *Uncertainties:* Construction contains many inherent risks. Most construction activities contain uncertainty, particularly when associated with the ground and/or other unforeseen conditions.
- *High Value:* Most construction activities, other than minor maintenance operations, incur considerable cost. Most clients finance this cost through capital investment schemes over and above their operational costs. Often third party finance is required to pay for the outlay, to which there will be additional terms and drivers attached in order provide the external lender with the necessary pay-back required. This will place additional demands on the client when considering how to procure construction works effectively.
- *Strategic Importance:* Finally, many construction works (particularly new build and refurbishment schemes) are required as part of the strategic development of the client's primary business. This may mean there are requirements on the construction works to be completed to a specific time or budget, *inter alia*. Although construction of the asset will be a support service to the client's primary business, the creation (or sustenance) of that asset will, nevertheless, be of some relative criticality to the main business (i.e. it will be important for the on-going success of the client's business in an entirely different supply chain). This will (or should) effect the manner in which the construction is procured (again, refer to Cox and Townsend[7]).

Nature of supply

Many of the following properties have already been discussed in Chapter 1. Needless to say, the nature of the supply-base is a product of the nature of demand and the way in which clients have pursued their procurement policies to date.

- *Fragmented Structure:* Construction is important to the national economy: it is the UK's largest industrial sector, three times larger than agriculture and employing 1.4 million direct employees. Of the 350,000 companies registered in the UK, 51% (or 180,000) are construction companies[8]. However, with an output of approximately £50 billion per annum, construction represents just 9% of the nation's Gross Domestic Product. Needless to say these figures demonstrate the degree of fragmentation within the supply-base. As a result, there are many cost-inefficiencies associated with contracting works (in particular transaction costs). In Chapter 1 it was noted that only 5% of construction firms (by number) have more than seven employees and yet they conduct nearly 70% of the national workload (by value).

- *Low Barriers to Entry:* One of the reasons for the high degree of fragmentation is the number of self-employed operatives and small locally-based family-run firms. Any organisation may commission construction works (and thus become a 'client'), irrespective of their knowledge, understanding and/or competence. Similarly, anyone owning a van or a wheelbarrow can form their own construction firm, irrespective of qualification and/or competence. As such there are extremely low barriers to market entry. While this remains the case, the supply-base will continue to be fragmented with highly variable standards of quality and competence.

- *Hyper-Competition:* With low barriers to entry and an over-supply of rival firms, the competitive forces on the supply-base are extraordinarily high. In many ways, this industry offers the closest resemblance to the economist's perfect market albeit that, somehow, the supply base is yet to reflect Adam Smith's theory of wealth generation.

- *Low Margins:* Between 1992 and 1997 successful construction firms have recovered a net profit of 1% their annual turnovers. The high level of competition has meant that some contractors would sooner win tenders on the basis of loss-leaders, than tender the actual cost of the works. This practice (combined with the fact that many clients accept the lowest priced tender, regardless of the true cost of the works) has placed extreme pressure on site staff to recover revenue through on-site variations, contractual claims and disputes.

- *Many Insolvencies:* With these practices prevalent, there have been a considerable number of insolvencies (more than 16,000 between 1990 and 1996). Moreover the threat of insolvency has hung over many contractors for years. Rather than make them entrepreneurial risk-takers, this climate has bred outwardly adversarial short-term attitudes throughout the supply-base.

- *Adversarial Culture:* This is examined in greater detail in Chapter 12 which considers the level of conflict in the industry and concludes that, at any point in time, approximately 10% of business in construction is in dispute. This has the result of spawning a parasitic sub-industry of claims consultants and construction lawyers paid for and developed by the industry's clients.

- *Lack of Training:* As a result of the fluctuating and uncertain workload, contractors are reluctant to employ large numbers of employees and maintain high levels of training and staff development. Instead they prefer to contract work activities as and when they are required, according to the current level of demand. The result is that few organisations are able to offer their workforce sustained training programmes and the competence level of the industry is severely affected. It is not that all contractors and their suppliers are incompetent; they simply lack the guarantees from clients to know their future business and training requirements.

The mis-match of supply and demand

Clearly there is a mis-match between supply and demand. Past studies of the industry have focused on the supply market and sought ways in which the industry's level of competence and competitiveness can be enhanced. This, in the authors' view, is the wrong approach. As mentioned before, the nature of the supply-base is a product of the way in which clients choose to procure their works. While the workload demanded by clients continues to be tendered, in piecemeal fashion, on the basis of lowest-price wins and no guarantees or incentives for the future, clients can expect to have short-term attitudes and behaviour in response from contractors.

It is this dysfunctional match between demand and supply that lies at the heart of the industry's problems (see Figure 3.3).

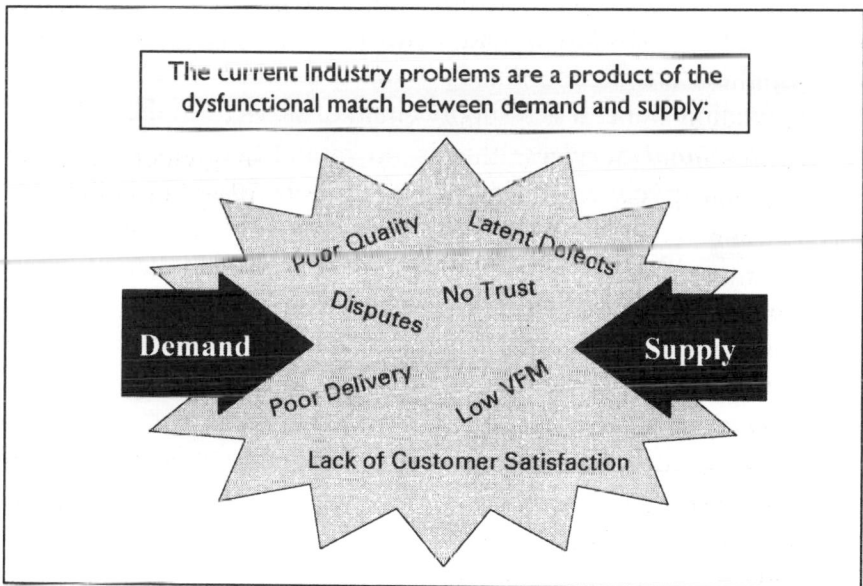

The current industry problems are a product of the dysfunctional match between demand and supply:

Poor Quality   Latent Defects
Disputes   No Trust
Demand   Supply
Poor Delivery   Low VFM
Lack of Customer Satisfaction

Figure 3.3: The Mis-Match of Supply and Demand.

Supply Chain Characteristics

Having examined the nature of supply and demand and identified that many of the current problems are a direct result of the mis-match that exists between them, it is now possible to have a better understanding of the industry's supply chain(s). Once the supply chain characteristics have been determined it is possible to codify an appropriate and effective approach to procurement in order to achieve the strategic goals of the business.

Is there one supply chain?

Clearly there is more than one supply chain in construction. It is no monolithic industry and commonly used terms, such as 'design and build', give an obvious indication of some of the differences that exist in terms of supply. Indeed there are numerous supply chains throughout the industry, although these can be codified under three generic headings:

- **Building and Assembly:** this category of supply chains is most commonly associated with the 'doing' part of construction activities and includes the main contractors, builders and specialist trades contractors. The activities involve fitting, assembly, installation and general on-site labouring. These suppliers are predominantly trading on their skills-base and reputation alone; it is a supply chain of *services* to the industry.

- **Professional Services:** this is no monolithic sub-division of the industry either; indeed it is even more fragmented than the building and assembly supply chains with a multitude of activities including planning, design, financial and legal control, project management, supervision, etc. These suppliers trade on their skills-base, intellectual property and reputation; again they supply *services* to the industry.

- **Materials and Products:** the third generic category of supply chains comprise the physical supply of goods and plant to form the constructed asset. Again there are sub-divisions, including the supply of raw materials (concrete, steel, stone, timber, etc.), the supply of products (such as pumps, windows, damp-proof course, etc.) and the supply of hired plant (cranes, excavators, pumps, scaffolding, etc.). These suppliers trade on the *functionality* of their goods and products; they comprise a supply chain of goods *and* services to the industry.

Structure of the Industry

The purpose of identifying these different categories of supply is not to confuse, it is simply to recognise that the structure of the industry is complex. Each supply chain has different properties and characteristics and yet, despite these differences, the client usually only sees one supply-offering when approaching the market (see Figure 3.4).

This figure illustrates the three generic construction supply chains forming one supply-offering to the client. The client is not always the same entity as the end-users ('consumers') of the constructed asset and this is illustrated by the various groupings positioned next to the client in Figure 3.4. This illustration also shows that the client has a position within its own supply chain. This supply chain constitutes its primary business (the *primary supply chain*) and will have its own business goals and drivers. Thus, for the client, construction is only a support supply chain.

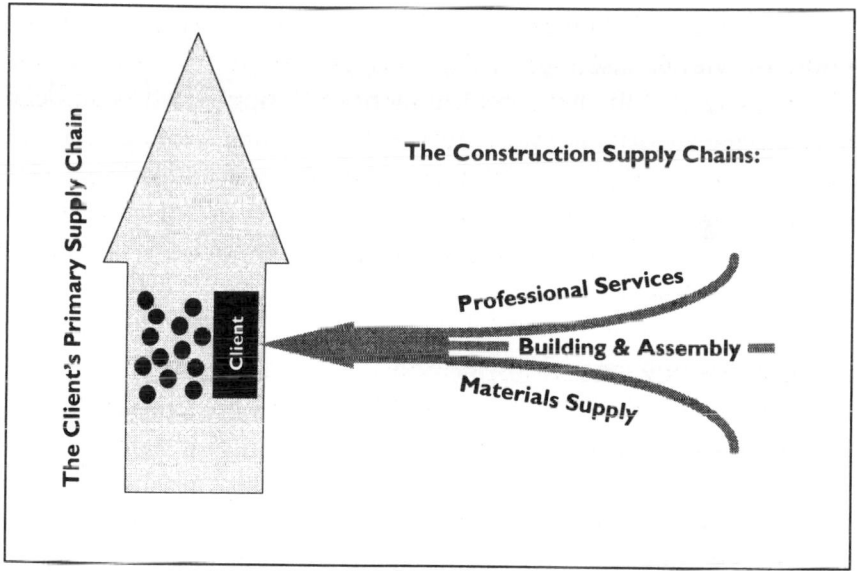

Figure 3.4: The Construction Supply Chains.

The problem is: from whom does the construction supply-offering come? In their research, the authors regularly ask clients' procurement staff who they procure their construction from. There are many answers including: architects, project managers, main contractors, developers, lawyers, consultants, surveyors, engineers, etc. Despite the obvious differences, they all seem to offer the same service: i.e. that of 'construction'. The significance of this, however, is that behind the initial supply-offering there lies several different industries within the one. These industries are dissimilar and may need to be managed in dissimilar ways, according whatever is appropriate for the specific circumstances of the supply and demand.

Effective Supply Chain Management

Thus, to return to the examination of supply chain management and its relevance to the construction industry, it will be clear to the reader by now that there is no blanket approach that can be advocated. Despite this statement of common sense, there are some who still ardently believe that Japanese automotive industry practices should be applied throughout the construction industry[9].

Clearly this is a nonsense. This book takes a totally different point of view, based on what is *appropriate* for the specific circumstances of the business transaction[10]. Supply chains *do* need to be managed; but not through the blind application of management practices which have been appropriate under some circumstances, but which may be wholly inappropriate under others; such thinking would constitute the adoption of business 'fads'.

A contingency-based approach

It is clear, therefore, that there will be different approaches and practices which can be adopted when contracting for works. In the three chapters thusfar, the authors have developed an argument to consider first what the intention (purpose) is of the business transaction and then to proceed to identify what is most appropriate to deliver this end-goal. Thus it is contended that a 'fit-for-purpose' approach is required[11]: there are no blanket answers.

The key to knowing what is fit-for-purpose is to develop an understanding of the contingent circumstances of the environment in which we are contracting, together with the knowledge of what, out of all possible techniques that could be applied, is most appropriate to serving our need. This is the *contracting competence* that was introduced in Chapter 2.

The remainder of this book is dedicated to the development of this concept of *appropriateness* in contracting for construction works. Each of the common contracting methods and their respective standard forms of contract are considered in turn. Through each examination, the following questions have been raised:

- who or what does this contracting practice serve?
- when is it appropriate to be used?
- what are the allocations of power and risk within the contract?

Through these discussions, it is hoped that a better understanding of appropriateness is developed and thus this book offers the reader a greater degree of *contracting competence*.

Further Reading

Few publications examine the application of supply chain management practices in construction. This book predominantly focuses on the contractual nature of the relationship between parties within the construction supply chain(s). However it has been published with a parallel text which examines procurement management throughout construction and the contribution it can make to the strategic intent of the organisation. For a fuller discussion on procurement issues in the construction industry, reference to this text is recommended:

Cox A. & M. Townsend (1998) *Strategic Procurement in Construction*, Thomas Telford Ltd, London.

Thompson I. & A. Cox (1998) *The Contract Selection Toolkit: Practical guidance for the construction industry*, Earlsgate Press, Boston, UK.

For a broader discussion on the merits of various supply chain and business management concepts and practices in general, the following texts are recommended:

Cox A. (1997) *Business Success: A Way of Thinking About Strategy, Critical Supply Chain Assets and Operational best Practice*, Earlsgate Press, Boston, UK.

Cox A. & P. Hines (1997) *Advanced Supply Management: The Best Practice Debate*, Earlsgate Press, Boston, UK.

Chapter Notes

1. Harland C. M. (1996) 'Supply Chain Management: Relationships, Chains and Networks' *British Journal of Management*, Vol. 7, Special Issue (March), pp. S63 - S80.
2. Cox A. (1997) *Business Success: A Way of Thinking About Strategy, Critical Supply Chain Assets and Operational Best Practice*, Earlsgate Press, Boston, p. 207.
3. Womack J. P, Jones D. T. & D. Roos (1990) *The Machine That Changed The World*, Rawson Associates, New York.
4. Forrester J. (1958) 'Industrial Dynamics: A Major Breakthrough for Decision Makers' *Harvard Business Review* (July/August) pp. 37 - 66.
5. Hamel G. & Prahalad C.K. (1994) *Competing for the Future*, HBS Press, Boston, MA.
6. The facts presented in this case study were first presented in *The Machine That Changed The World (op. cit.)*. The authors would also like to thank Professor Peter Hines for his contribution to this case study with data concerning Toyota Homes which was taken from Toyota's annual report and accounts.
7. Cox A. & M. Townsend (1998) *Strategic Procurement in Construction*, Thomas Telford Ltd, London.
8. Raftery J. (1991) *Principles of Building Economics*, Blackwell, Oxford.
9. Womack J. P. & D. T. Jones (1996) *Lean Thinking*, Simon & Schuster, New York, p. 291.
10. The concept of *appropriateness* has been developed by one of the authors as a result of 25 years study of business management practices in the political economy. Refer to: Cox A. (1997) *Business Success: A Way of Thinking About Strategy, Critical Supply Chain Assets and Operational best Practice*, Earlsgate Press, Boston, UK.
11. Refer to: Cox A. & I. Thompson (1997) 'Fit for Purpose Contractual Relations: Determining a Theoretical Framework for Construction Projects' *European Journal of Purchasing & Supply Management*, Vol. 3, No. 3, pp. 127 - 135.

Chapter 4

Introduction to Contracting Methods

Introduction

This chapter introduces the reader to current contracting practice in the construction industry and the various methods of contracting that currently exist. It is a descriptive summary which serves as a precursor to Part B which provides a more detailed analysis of each of these methods and their respective forms of contract.

By 'contracting method' is meant that process by which construction works are procured. Predominantly this refers to works that are bought from the external market, using a contract as the vehicle for the commercial transaction (as discussed in Chapter 2), although a brief section on 'internal' contracts has been included here.

The contracting method is referred to as a 'process' because this offers the simplest description of all the activities that are required in order to implement and deliver the completed construction works. Generically these activities will be the same for whatever construction works are required. However the order in which they are carried out and, moreover, by whom the activities are supplied will differ from project to project depending on the contingent circumstances of the business. Thus different contracting methods have evolved to order the construction process in a variety of ways so that the client sponsor may choose the optimal approach to contracting the works.

This is most clearly demonstrated in the design activities for construction and the discussion as to whether these are best conducted prior to the award of the works contract or after the contract award by the works contractor, as in the design and build option.

In many construction management texts, the contracting method is referred to as the contract strategy, contracting strategy and/or procurement strategy. This is not necessarily wrong, but it does give a blinkered and myopic view of strategy and of procurement. There is very little strategy involved in choosing between the types of process and, furthermore, at the corporate level of business, this decision is viewed as one of many other operational considerations that have to be made in the course of everyday business. The decision is important and it may effect the successful delivery of that physical asset to the client, but not every important decision is *strategic*.

Furthermore, as may have become apparent to the reader so far, the concept 'procurement' is seen here to be far more encompassing than is credited in current construction practices. Procurement is the process which comprises all activities required to deliver goods and services into direct ownership and/or use. Furthermore good procurement practice will account for all of the contingent variables in the commercial environment in order to deliver those goods and services in a way that maximises the business objectives of the organisation (or individual). The considerations required at this level are more complex and strategic than current practice would suggest. These include issues of outsourcing, the make/buy decision, optimal sourcing relationships and supply chain management, *inter alia*.

The main differences between contracting methods are *structural*, in that their respective forms of contract are structured in a number of alternative ways. This usually concerns the implementation of the design element of the works and the number and type of suppliers carrying out each element of the works. This will bear important consequences on the overall procurement of the works in terms of risk, liability and responsibility. It will also have an important bearing on the cost of the works and the way in which suppliers are reimbursed.

Other differences in contracting methods can be codified in terms of specification and in relational terms. For example performance contracts are substantially different to other standard forms of contract and the construction industry has much to thank for the innovations from the plant and process sector. Similarly, contracts can be distinguished by the way in which their reimbursement mechanisms are specified (as in the cost-reimbursable, admeasurement and lump sum methods).

'Partnering' and other relational contracting strategies provide an additional dimension to this classification of contracts in that they introduce transactional behaviour to the contracting methods.

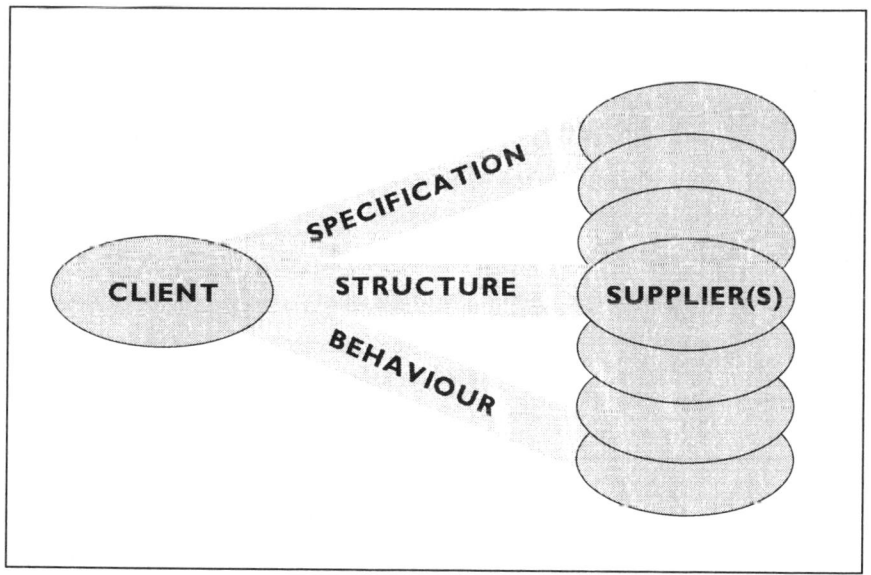

Figure 4.1: Three Dimensions of Contracting Strategy.

Thus contracting strategies are subject to a three-way cut (as shown in Figure 4.1). Each dimension needs to be considered in turn, in order to place the most appropriate contract with the most appropriate supply-offering and thus appropriate the business objectives from the transaction. The resultant contracting method, built around a fit-for-purpose relationship and form of contract, provides a structure of power that over-determines the way in which the business transaction is conducted.

As noted in the introduction, the primary analysis of this book is to consider the structure of various standard forms of contract in order to assess the balance of power in the ensuing business transactions.

The Contracting Methods

The following contracting methods are discussed with comments on their popularity and claims. The purpose of the chapter is to give the reader a brief insight into each of the methods and their use, merits and dangers. It is meant for use as a point of comparison and not as a robust selection guide. Most of the following methods and their respective standard forms of contract are examined in closer detail in Part B of this book. The contracting methods introduced in this section comprise:

- Sequential contracting;
- Design and build contracting;
- Performance contracting;
- Partial design 'sequential' contracting;
- Minor works contracting;
- Management contracting;
- Serial contracting and framework agreements;
- Extended arm contracting;
- Partnering;
- Relational contracting;
- Internal contracting.

Sequential contracting

Sequential contracting is the most commonly used approach to the procurement of construction projects (new build, refurbishment and some renewals works). Typically it is referred to as the 'traditional' method of contracting and/or the 'traditional' procurement approach. Both these terms are misleading, unhelpful and, moreover, inaccurate. Much of the literature in purchasing and/or supply management refers to a traditional approach which suggests that the old ways are not necessarily yielding the most advantage from commercial transactions; it is more akin to a clerical buying process than anything else; but this is not what is meant by traditional procurement in the construction management literature. Moreover, it is questionable whether this type of procurement really is the 'traditional' way. In the days of major infrastructure investment, clients (usually public bodies) procured their works through large construction firms on the similar line to a management contractor today, albeit that many of the activities were vertically integrated. Thus we have adopted an alternative term to give a better reflection of the contracting process involved in this method.

Sequential contracting refers to the process whereby clients undertake the design activities separately from the works activities. The design is completed prior to the award of the works contract, either by the client or through a third party professional designer; this is then used as part of the tender enquiry documentation for the works contractor to price. The process of project design followed by works implementation occurs *sequentially*; hence the adoption of our terminology.

The main advantage of this process is the retention of control on the specification of design and quality of finished works. It means that the client carries the risk that the works can be implemented to the requirements of the design (i.e. they are 'buildable'), which is especially appropriate if the design requires specialist input which only the client will have knowledge of. It also means that, in theory, the price for the implementation of the works will be cheaper (since the contractor does not have add a risk premium for the effectiveness of the design).

The disadvantage of sequential contracting lies in the process itself. By its nature it is time-consuming; the design must be complete before the works contract can be awarded. Other methods allow for greater integration of the activities which compresses the total duration of the project. Furthermore, there is little time or cost certainty with the sequential method; the client has accepted the risks of the design and its ability to be implemented. Where amendments and/or variations are required, there will be a consequential effect on the contract sum and possibly the contract duration too.

Finally one other disadvantage of fragmenting the activities of design and construction is the additional transaction costs involved in another external contract. Of course, if the design is carried out in-house, this is not a concern.

In Chapter 5, sequential contracting is examined in greater detail. The chapter considers the following standard forms of contract, which are contracts for the works elements only and not the design contracts:

- ICE Conditions of Contract, 6th Edition;
- JCT'80 Standard Form of Building Contract;
- FIDIC 4th Edition.

Design and build contracting

Design and build contracting is the obvious counterpoint to sequential contracting. In design and build, the contractor takes on *single point responsibility* to design and deliver the works according to the client's requirements. The method suggests that significant advantages can be gained from contracting the design works from the same supplier contracted to implement the works. These advantages include greater certainty for the client in terms of time and cost of delivery; the works are more likely to be 'buildable' and delivered to the client more quickly.

The assumption in design and build is that the client is capable of articulating its requirements for the contractor to develop a design and subsequent works. If this is not the case, the client will need to commission a professional project management firm, or similar, to specify its requirements or run the risk of having a construction project that does meet its expectations.

Chapter 6 examines the design and build process in greater detail, including the following standard forms of contract and some of their provisions:

- ICE Design and Construct Conditions of Contract;
- JCT'81 Standard Form of Building Contract with Contractor's Design;

- IChemE Red Book Model Form of Contract.

Performance contracting

This is a contracting method that has come from the chemical plant and process industries, where the completed works are required to carry out a specific function and, within certain bounds, it does not matter what the works are like. As such, the client specifies the *performance* required of the completed works (as well as other criteria such as design life and serviceability, etc.) and the works contractor accepts responsibility for delivering the completed works in such a way as to provide it. The performance may be specified in any number of ways, depending on the type of asset to be constructed. For example, it may specify batch requirements, production capacities and/or consent standards. There will also be requirements to manage interfaces with other operational assets, as well as for testing and commissioning.

In effect performance contracting is very similar to design and build contracting, except the client is less concerned with the appearance of the design and more concerned with the functionality and performance of the finished works; it is not an architect's approach to construction!

Forms of performance contract are commonly available in the industry through the Institution of Chemical Engineers who have published two main contracts: the Red Book (which is a lump sum design and build contract discussed in Chapter 6) and the Green Book (which is discussed in the more detailed section on performance contracting in Chapter 9).

Partial design sequential contracting

This is a 'hybrid' version of contracting methods where the client initiates the design prior to the works contract being awarded but leaves part of that design for the works contractor to complete. This has the advantage of accommodating the specific needs of the project and capabilities of the supply market, although the design interface can be difficult to manage. In practice it is easier to allocate different sections of the design to the different parties, rather than handing over a partially complete design. This is not only better management of the actual design, it helps prevent issues of unresolved design liability occurring.

This is considered in greater detail in Chapter 5, which considers the ICE 6th Edition Conditions of Contract and its provisions for partial design. The JCT'80 can be modified with supplementary clauses to provide for partial design too.

Minor works contracting

This is not really a separate contracting method from any of the previously mentioned processes. However, several construction contracts have been especially drafted for smaller works. In this context, 'minor' refers to works of smaller value. The process is not any different, nor are the risks involved, except in magnitude. Nevertheless there is an argument that smaller works do not justify the same administrative burden of a full set of conditions of contract and therefore can be 'thinned' down accordingly.

There is an obvious assumption here that suggests that smaller works are of less importance to the business and therefore require less legal protection. If this is the case, the unsuspecting client could place the project or, more importantly, elements of the business at considerable risk. This is discussed in greater detail in Chapter 7 where following contracts are examined:

- ICE Minor Works Conditions of Contract, 2nd Edition;
- JCT Intermediate Form of Contract 1984;
- JCT MW'80 Standard Form of Building Contract for Minor Works.

Management contracting

There are two types of management contracting both of which are used for larger works projects that usually require the specialist services of several different types of trades contractors. The works are packaged into various types of works contract and let separately to contractors of differing disciplines. The construction requires a multi-disciplinary approach, of which the process is managed by a 'management contractor'. The two types of management contract are either *management contracting* or *construction management*.

The differences have important legal consequences for the client. In management contracting, the management contractor accepts a single point of responsibility to the client and separately sub-lets the various works contracts (the client is not privy to these agreements). In construction management the management contractor still manages the process but acts as the client's agent for each of the works packages which are let between the client and the individual trades contractors (the client is privy to these contracts).

The relative merits of each of these processes is considered in Chapter 9, which also specifically examines the JCT MC/87 management contract.

Serial contracting and framework agreements

Serial contracting is not a revelation to those in construction procurement, although it has only become of increasing popularity towards the late 1990s. Its rationale is that the consolidation of expenditure will increase a client's leverage in the market place and provide a more attractive workload for suppliers. Quite simply it operates by stringing together a series of works contracts, so that the same supplier is retained over a longer time frame. For the contractor this means the potential of a steady workload which does not have to be tendered for and 'won', while for the client this means the potential of reduced rates and increased productivity (see Case Study 2 overleaf). In essence it moves away from the arms-length basis of construction projects and introduces a more collaborative approach. The process is not difficult to manage and it does not contravene EU procurement directives; for the client who is able to offer a regular workload, it is an eminently sensible approach to construction procurement.

Framework agreements are similar to serial contracts, except that they operate for a given time period rather than a specific number of contracts. They are also called term contracts and/or call-off contracts. The idea is very similar to serial contracting and it produces similar cost and productivity advantages. The mechanism operates at two distinct levels: the framework agreement operates for the term of the appointment and stipulates the general conditions of engagement. However this rarely contains any orders or guarantees of work; the contractor is asked to price a notional list of work activities which he may, or may not, be required to conduct. Within the term of the appointment, the client will call-off individual work activities (or 'episodes') with the use of purchase orders, or similar types of documentation. Each work episode is conducted as a separate contract, albeit on preferred terms of reduced rates (for consolidated spend). The client will often be able to secure quicker response times, since there is no tendering process to go through. For framework agreements with local suppliers, or at the same site, the mobilisation time is likely to be quicker too.

Case Study 2: Tesco Stores Limited

The retail giant, Tesco Stores Limited, is one client that began to adopt this process in the mid-1990s. Tesco comprise the largest private building client in the UK with an annual expenditure of approximately £800 million in construction, of which £200 million comprises new building works. It owns just under 600 stores and has a programme of development of 20 new stores every year. Traditionally every new development was procured by individual competitive tender through the sequential contracting route. However more recently, Tesco has realised the market leverage that it can gain by pooling building contracts. Not only does it reduce transaction costs and management on-costs, but suppliers have the incentive to offer better terms to Tesco. For example: one supplier who was given a series of three similar contracts managed to save three weeks on the second contract and five weeks on the third contract, just from the learning curve advantage gained by the repeat process.

Extended arm contracting (EAC)

The EAC method has had an unhappy and unsuccessful initiation in the construction industry. Essentially it is another form of management contract, but where the management contractor is replaced by a 'core team' established by the client. The core team usually comprises project management staff from a contracting firm and, on occasions, some of the client's own staff too. The team is retained by the client to act as an 'in-house' management contractor (the 'extended arm') for a specified term of engagement. In this time the core team works with the client to develop works projects for implementation and sets up an manages works contracts between the EAC contractor and the client.

The system has been tried in several large client organisations with considerable portfolios of construction expenditure (such as Thames Water, Railtrack and London Underground). It encourages large clients to use EAC as a means of having a management resource under its control without being on its permanent payroll. However, there is no guarantee that the EAC contractor will be given any works to implement in the course of the engagement and this has led to a conflict of interests in some of the organisations mentioned.

Partnering

Partnering (also called 'partnership sourcing') is the latest "fad" to reach construction. It has been endorsed by public figures such as Sir Michael Latham[1] and exploited by many consultants and academics alike. Before reaching construction it held the limelight in many business and management texts[2]. Many consider it to offer the panacea to the industry's ills, but unfortunately few are able to concisely say what the concept involves.

The definition adopted by the DOE's Construction Industry Board Working Group 12, which was charged with reporting on partnering[3], was lifted directly from the Reading University's *Trusting the Team* report: *"...a management approach used by two or more organisations to achieve specific business objectives by maximising the effectiveness of each other's resources."*[4]. Unfortunately this is so general and uninformative, that it is of little use. The definition offered by the European Construction Institute is not dissimilar[5].

In the early part of 1997, the authors of this book conducted a survey of the top 100 UK construction clients and the top 100 UK contracting organisations to consider the issues of conflict and disputes in construction. Unsurprisingly (in accordance with the industry's dominant paradigm) 85% of respondents agreed that partnering offered the greatest opportunity to reduce the level of Adversarialism and dispute in the industry. However when asked to define what respondents considered partnering to be, there was no singly agreed response (Figure 4.2 illustrates the variety of responses received). Clearly it means different things to different people, which is a major problem if parties to a 'partnering' contract cannot agree what is meant by the process.

What does partnering involve?	No. of survey respondents who agreed: (n = 77 respondents)
Mutual objectives	12
Working together	10
Benefit sharing	8
Risk sharing	7
Repeat business	7
Openness	6
Open book accounting	6
Team working	6

Figure 4.2: What the Industry Considers Partnering to Entail.
(Note: none of these figures is statistically relevant; what they do demonstrate is the inability of the industry to agree on what constitutes the partnering concept.)

Partnering seems to refer to a collaborative close-working relationship between the parties to a contract. Its mechanisms operate at different levels. For some, all partnering means is an opportunity to embrace 'team-working' and joint problem-solving instead of an adversarial approach to claims and dispute. This may occur for just one project (so called *project-partnering*[7]) or over a series of project (so-called *strategic partnering*[7] but, in reality, nothing more than the serial contracting method discussed earlier). Other more integrated levels of partnering include joint risk and reward sharing mechanisms, shared resources and open-book accounting, *inter alia*. As partnering has continued to be promoted many have become sceptical that it is nothing more than a marketing exercise offered by contractors who, in desperation for a steadier workload, will agree to any alternative way of working. To the promulgators of partnering this is evidence of the approach's success; however, as has been discussed elsewhere[8], partnering is only likely to work under certain defined and specific circumstances.

Relational contracting

The concept of relational contracting suggests that there are certain circumstances in business transactions where a contract is not necessarily required. In these situations there is a governing relationship between the client and contractor which is so dominant that the need for a contract becomes redundant; indeed the presence of a contract could become an inhibiting factor and has the effect of adding unnecessary transaction costs to the construction process.

It should be noted that these occasions will only apply when there are dominant relational structures in place to curb contractual opportunism. It is unlikely that this type of contracting will be able to succeed in many areas of the UK's construction market. However the principles behind relational contracting and the role of a governing relationship are very important strategic factors which have been discussed in Cox & Townsend[8].

Internal contracting

Any discussion on the current contracting methods available in the construction industry would be incomplete without a mention of internal contracting. Traditionally, many construction activities were vertically integrated and, even recently, many maintenance teams were contracted through 'in house' direct labour. These work-forces carried the stigma of being expensive and inefficient, compared to the external market. The UK Government's policy of Compulsory Competitive Tendering has the effect of outsourcing many of these operations and further fragmenting the supply market.

While there is insufficient space to develop an extensive critique of this and similar thinking among private sector companies, it is worth remarking on the fact that it does not follow at all that external sourcing from private sector construction specialists will always be more efficient or economic than in-house contracting on the basis of vertical integration. Since external suppliers exist to make a profit, if the level and regularity of construction work for any public or private sector organisation is of a sufficient magnitude, then it may be more appropriate to manage the activity in-house. This is a fundamental question for many public and private sector companies to address through effective make-buy methodology that focuses on the total cost of ownership, the criticality of the assets to be created and the scope for external supplier opportunism. This is just part of a strategic approach to procurement management in construction.

Types of construction work

For most of the contracting methods mentioned in the previous section, there has been a choice of standard form of contract. Most of these forms of contract have been developed by joint industry drafting committees. The result is a compromise of beliefs to form a 'fair and equitable' balance of risk and power in the contract terms. However some of these committees have been more successful than others. Case Study 3 describes the historical background and structural content of the Joint Contracts Tribunal (JCT). In his review of contractual and procurement practices, Sir Michael Latham was keen to re-structure both this drafting committee and the CCSJC (the ICE's equivalent drafting committee). The JCT committee has rarely represented the industry's views and this case study demonstrates the institutional fragmentation that exists in the building industry, which consequently lays a foundation of disagreement and dissatisfaction in the current standard forms of contract.

Apart from the protectionism of the various professional and trades institutions, there are some plausible reasons for separating building contracts (provided by the JCT) and civil engineering contracts (provided by the CCSJC). The works have different content which require different approaches in the apportionment of risk. For example civil engineering risks have a stronger association with earthworks, foundations and the natural environment. These works carry a more significant proportion of the unforeseen and the risks should be borne by the client. Building works, on the other hand, are more associated with structures and furnishings in the built environment. These works have a greater need for specific types of insurance to indemnify the client against third party interests from adjacent property owners and the like.

Case Study 3: The Joint Contracts Tribunal (JCT)

The traditions of the JCT building contracts extend back to the nineteenth century when independent trade bodies and building representatives drafted their own forms of contract for building works. The need for a form of contract was driven by the desire to cover all eventualities that may lead to risk exposure; that is to safeguard against the likelihood of additional costs or delays occurring (or whatever else the criterion of the building works may be). Prior to the introduction of a standard form, clients and their professional advisers would draft their own contracts to suit their perceived needs of the building works. Contractual relations were focused on the nature of the building works and thus became production-led and short-term (i.e. works were contracted on a one-off project-by-project basis). Essentially they were (and still are today) *reactive* mechanisms designed to apportion responsibility for actions and events as they are or were discovered. These bespoke forms varied greatly and were generally not well-received by the industry as they tended to protect the needs of the client by passing the risks along the supply chain. Similarly the courts took a dim view of the legality and fairness of some of these documents.

The first moves towards standardising the contracts came in 1870 when the building trades began to discuss the issue with the RIBA, however agreement could not be reached until 1903 when a settlement was achieved through the mediation of the (then) Institute of Builders and a standard form was published. Despite revision in 1909, the members of the National Federation of Building Trades were still unhappy with this agreement and it was rescinded in 1919. The Builders published their own standard Building Code whose terms in turn were refused by the architects.

Throughout the 1920s the discussions between architects, surveyors and builders continued and a joint drafting committee was established to agree on a standard form. The committee produced a draft contract in 1928 which was endorsed by the RIBA Council but rejected by the affiliated Architectural Societies. In 1931 negotiations were finally settled and the 1931 Form of Contract was published. It was accepted by all the main parties and became known as the 'RIBA Form'.

Concurrently, a new committee was appointed by the RIBA to keep the form up-to-date with modern practice and free of any difficulties which may arise in connection with its use. This committee was called the 'Joint Contracts Tribunal' and is the patriarch of the contemporary JCT contracts.

The formation of the JCT did little to ease industry tensions over the nature of this contract and many clients continued to amend the clauses to suit their own needs. Particular concern came from local authorities who, despite the publication of a revision adapted for use by local authorities, continued to alter the form's terms.

The four Local Authority Associations were unable to agree with the JCT that the contract was fair and reasonable and, in an attempt to resolve this, the JCT was reconstituted to include these Associations. In February 1956, it invited comments from the industry on the contract and, in response, received a total of 1400 points of issue. This eventually led to the publication of the 1963 Edition of the Standard Form (the JCT'63) in four forms: those with or without quantities and those for private or local authority use. This was later revised to become the JCT Standard Form of Building Contract 1980 (the JCT'80).

However the conflict surrounding the structure of the JCT and its industry representation has not stopped here. In the course of 1997, the *Building* magazine reported the disagreements between contractors and subcontractors surrounding the restructuring of the JCT to become a limited liability company. The JCT looks set to continue being a battlefield for power in the industry.

Source: JCT [9]

The differences are tenuous and, under the challenge of new 'non-institutionalised' contracts, they appear weak. Both the NEC and GC/Works/1 require the use of a project manager (no instituted professionals are mentioned; be they engineers, architects or surveyors) and this is appropriately so. The issue should not be whether the contract manager is chartered or not, it should be whether he/she is competent to carry out the required task. In this view, the continuing support for ICE and/or JCT contracts clearly diminishes.

Conclusion

This brief introduction to the contracting methods and the development of the standard forms of contract has served to indicate that the industry is contracting its business by means of historical artefacts. Some of these artefacts will be appropriate under certain circumstances, whereas others are outdated and inappropriate products of institutional in-fighting.

The tensions that exist between the various industry sectors are more than mere professional snobbery; they constitute part of man's eternal conflict for the control of scarce resources. Each form of contract offers a structure of power and risk apportionment to the business transaction which shapes the way business is conducted. From this structure suppliers will be required to provide construction services and, in return, appropriate money for themselves. The structure of power in each contract therefore becomes a determining factor in the business exchange.

The following chapters (in Parts B and C) proceed to examine each of these standard forms of contract and their inherent structure of power. From this study, it is hoped that the reader will receive an emergent understanding of what constitutes 'fitness for purpose' for contracting for business success in the construction industry.

Chapter Notes

1. Latham M. (1994) *Constructing the Team: Final Report of the Government/Industry Review of Procurement and Contractual Arrangements in the UK Construction Industry*, HMSO, London.
2. See for example: Lamming R. (1993) *Beyond Partnership Sourcing*, Prentice Hall, Hemel Hempstead; and MacBeth D. & N. Ferguson (1994) *Partnership Sourcing*, FT Pitman Press, London.
3. Construction Industry Board, (1997) *Partnering in the Team: a report by Working Group 12 of the Construction Industry Board*, Thomas Telford, London.
4. Bennett J. & S. Jayes (1995) *Trusting the Team: the best practice guide to partnering in construction*, Centre for Strategic Studies in Construction, Reading.
5. Cox A. & M. Townsend (1998) *Strategic Procurement in Construction*, Thomas Telford, London.
6. Cox A. (1997) *Business Success: A way of thinking about strategy, critical supply chain assets and operational best practice*, Earlsgate Press, Boston.
7. The use of these terms are taken from the Reading report *Trusting the Team (Op. cit.)* and are not those of the authors.
8. Refer to: Cox A. & M. Townsend (1998) *Strategic Procurement in Construction*, Thomas Telford, London.

9. Joint Contracts Tribunal (1990) *The Use of Standard Forms of Building Contract*, RIBA Publications Ltd, London.

Part B

Forms of Contract

Chapter 5

Sequential Contracts

Introduction

In Chapter 4, Sequential Contracts were introduced as contracts which cover the works-implementation part of the construction process, where the design and project development processes have been separated from the works elements and carried out pre-tender. This contracting method is commonly referred to as the 'traditional' method; however, as outlined briefly in Chapter 4, this may be an inaccurate and misleading term. Since the constituent elements of the construction process are carried out sequentially in this procurement method (see Figure 5.1), it follows that 'sequential contracting' provides a more informed nomenclature.

Thus, sequential contracts only cover the physical construction of the works and none of the pre-construction elements; there has been a *separation of roles* between designer and constructor. Sequential contracts are therefore built around the principle of *contractual compliance*. The works contractor agrees to construct the works in accordance with the design and specification which has been prescribed by the client. Thus the client is required to prepare the design details prior to the award of the contract and, then, to employ a *supervising agent* to oversee the mode and manner of construction and ensure compliance with the specified details of the design. This agent has been supported by the professional institutions as the Engineer or Architect in the principle of *tri-lateral governance*; the agent is a creature of the contract and yet also a stranger to it. The engineers and architects have been given special powers under these contracts which, historically made them very powerful players in the construction process. Until recently there were few who questioned the professional standing of these people. This is reflected in the contracts which require the client's agent to have both *impartiality* to the contract and yet *immunity* from its provisions.

These general principles (the separation of roles, contractual compliance, the supervising agent, tri-lateral governance and the impartiality and immunity of the client's agent) are considered in more detail in the following section.

The chapter then proceeds to consider the business principles behind three of the most common standard forms of sequential contract:

- the ICE 6th Edition Conditions of Contract;
- the FIDIC 4th Edition; and
- the JCT Standard Form of Building Contract 1980.

It should also be noted that there are other standard forms of contract which act as sequential contracts. These have been discussed elsewhere in this book and include the NEC (see Chapter 8) and the Government Contracts: DEFCON 2000 and GC/Works/1 (see Chapter 12).

General Principles

Separation of Roles

The first principle is one that has already been covered in the introduction to sequential contracting and that is the separation of roles between designer and constructor. This contrasts with the design and build method of contracting where the contracted supplier is responsible for both main elements of the construction process. The effect of separating the roles has several consequences on the finished works and the commercial aspects of the transaction. Figure 5.1 indicates the sequential process; from which it is evident that few activities can be carried out in parallel. Thus the sequential contracting method is considered to be one of the most time consuming approaches to construction procurement. This is a trade-off for the control that this method gives over quality.

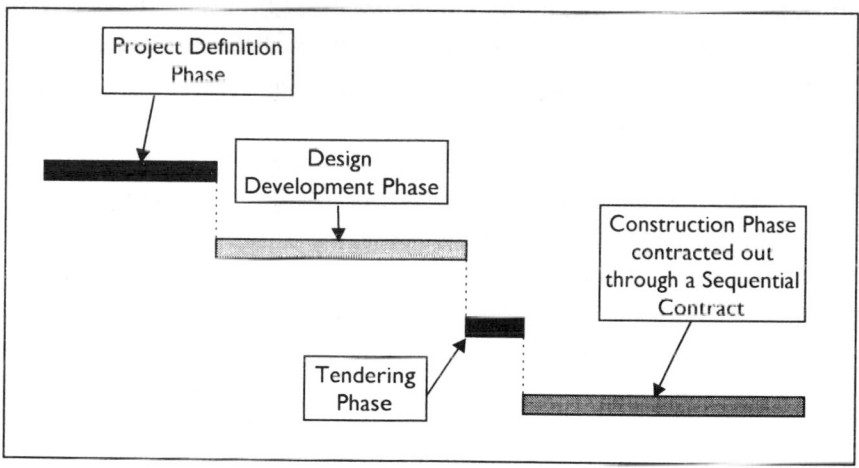

Figure 5.1: The Sequential Construction Process.

The issue of separate roles has other consequences too. The client is not taking full advantage of the main contractor's capabilities as a designer and, in retaining control of the specification and the design layout, the client is also accepting the risk of buildability. In the event that the design is not physically possible to construct, or that changes are required to the design in the course of constructing it, the client must pay for the costs of changes. In theory, these costs are based around the competitively tendered prices originally submitted by the contractor but, in practice, many changes result in additional sums of money being paid on a cost-plus basis. On this basis, the client has lost control of the costs of changes.

Other potential factors associated with the separation of roles brought about by sequential contracting include increases in transaction costs (as each element of the construction process is competitively tendered) and prime costs (as the client is preparing the design and may not specify as economically-advantageous details as would a commercial contractor).

Contractual Compliance

The main advantage of separating the roles of designer and constructor is in the retention of control that the client has over the specification of quality. This can be expressed in terms of the materials used and the mode or manner of the construction. For some clients this control will be essential, as they may be the only organisation with specific knowledge associated with the contracting of the works, such as works on an already operational site or one that is environmentally sensitive, *inter alia*. Similarly, the control on the specification may be desirable where the works are to be of considerable aesthetic value.

The issue of control of the design and specification becomes relevant in the course of implementing the works, as the works contractor is require to comply with that which is specified. The previous section has already discussed the effect this has on risk; the client assumes the risk of quality and, in the case of good workmanship, all works defects.

This has the potential to introduce conflict into the process. The client is specifying what the contractor should do in a one-way process. The contractor is required to comply, whether or not there is a better way to construct the works. It is a non-collaborative approach to constructing works, which assumes the specifier and designer of the works knows best and does not need the benefit of the constructor's experience and competence. Furthermore, supervising compliance becomes, in effect, a trial of the contractor's competence and workmanship. This inevitably leads to potential conflict surrounding questions of compliance and/or sound workmanship.

Supervising Agent

The issues of compliance and quality of workmanship in sequential contracts are resolved using an independent third party to supervise the construction process and preside over commercial or technical matters. This is the position of the 'professional'; the Engineer (in the case of ICE contracts) or the Architect and Quantity Surveyor (in the case of JCT contracts). The agent is given specific powers under the contract to monitor progress, check compliance and, if necessary, instruct remedies where the works do not comply with that specified.

Tri-lateral Governance

The presence of an independent third party agent introduces a specific dynamic to the contract which is referred to as *tri-lateral governance*. The third party governs the contract but is not privy to it. The principle of *privity of contract* has specific legal meaning that simply suggests that only the parties which agree to the contract are bound by its terms. This has clear ramifications on sequential construction contracts, as it means that the third party agent is not bound by the contract's terms although acting within them. As one commentary puts it, they are both a stranger to the contract and a creature of it[1]. This is best expressed in a simple illustration (Figure 5.2):

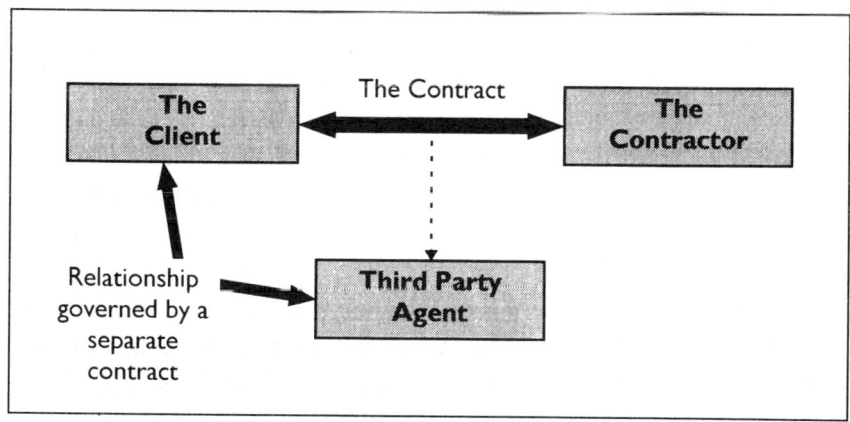

Figure 5.2 The Principle of Tri-lateral Governance.

Figure 5.2 demonstrates that the contract to construct the works is between the two principal parties: the client and the contractor. However the principle of tri-lateral governance introduces the third party agent to the mechanism of the contract, even though they are not privy to it.

The third party has been called an agent because in effect that is exactly what they are! They are employed under a separate agreement with the client to act in an independent supervisory capacity. It is important to understand that the agent's contractual duties are quite separate from the main works contract.

Impartiality of the Agent

The issues of tri-lateral governance and the role of the third party agent raise important questions of partiality. The agent is required to act independently and yet he is contracted under the payroll of the client. Furthermore, most Engineers and/or Architects have been involved in the preparation of the design details and contract documents and thus carry vested interests to ensure the details are complied with and that their adequacy is not challenged. An objective stance is difficult and unlikely to be assumed. Thus, as many contractors have suggested, the agent is not impartial and, therefore, cannot be independent. Their powers under the contract are wide-ranging and yet cannot be used without favour tending towards the client and his professional agent. This is especially so in an industry such as construction where opportunism is rife, even among the professionals.

Immunity of the Agent

Although the works contract effectively defines the terms under which the third party agent may act, it has already been established that the agent is not party to the contract. Thus, despite possessing influence and control over many aspects of the works, the agent is not liable for his/her actions under it. As such the agent has 'immunity' from the contract.

Both the Engineer and the Architect have no contractual liabilities to the Contractor and the doctrine of privity of contract prevents them from being sued for wrongful actions under the works contract by either Client or Contractor. Remedy can only be sought:

•   by the Client through their own separate terms of engagement;

- or, by either party for negligence in tort.

Interestingly, the latter recourse is unlikely as, in most cases, it would result in putting one party to the works contract in conflict with the other, when there are already contractual mechanisms in place within the main works contract to remedy this.

It is remarkable that the normal terms of engagement for a third party agent do not attempt, other than in a general sense, to define the limits of responsibility and liability of the third party. There is no ICE contract for the engagement of consulting engineers, for example. Thus the professional agent is engaged either through a simple exchange of letters (offer and acceptance) or through a contract for professional services such as the Association of Consulting Engineers' Conditions of Engagement or the NEC Professional Services Contract

In the absence of any express terms, statutory law prevails and *reasonable* skill, care and diligence in the discharge of duties is all that is required to be exercised within a *reasonable* time (as prescribed by The Supply of Goods and Services Act 1982).

Thus the client assigns considerable powers to a professional third party to act as an agent and govern the works contract; yet despite their considerable weight, few clients require those responsibilities to be supported by terms of liability and backed by the force of contractual law.

ICE 6th Edition Conditions of Contract

The principles of the ICE 6th Edition Conditions of Contract (ICE 6th, hereafter) are strongly embedded in the traditions of the Engineer and the contract is drafted around this central figure. The role of the Engineer has already been discussed and he is given wide-ranging powers under the contract (see Figure 5.3). His principal role is to ensure *contractual compliance*: that the Contractor shall comply with the express requirements of the contract and is subsequently paid his due.

The contract is an admeasure ('measure and value') contract and thus a schedule of rates for the completion of individual work items forms the basis of the tender and the award of the contract. *Thus, at the point of award, neither party actually knows how much the total works will cost the client.* The individual work items are based on the rules of measurement detailed in CESMM2 (the Civil Engineering Standard Method of Measurement, 2nd Edition, 1985); as prescribed under clause 57, although CESMM2 has now been superseded by a 3rd edition. For the sake of convenience, the sum of the rates is considered as a tender total, but this is liable to significant change as the works progress and the value of work done is assessed on an admeasurement basis. This provides considerable flexibility in the event of changes to the nature of the works. The variations (either instructed under clause 51, or merely in the quantity of work under clause 56) can be assessed on the basis of the rates and thus financial control is preserved in the light of changing requirements.

Thus the Engineer oversees the construction and completion of the works in the mode and manner executed by the Contractor. The Contractor is not responsible for the prescription of the works; this is supplied in advance by the Employer. However the Contractor is responsible for *how* the works are completed, which is apt since this should be one of his key competencies.

- Appointment of assistants and delegation of their authorities;
- Consent to sub-letting the Contract;
- Correction of ambiguities, issue of drawings and revision of information;
- Approval of the works programme;
- Removal of the Contractor's personnel from the site;
- Instructions for exploratory investigations;
- Instructions for tests and/or investigations on materials and workmanship;
- Removal and/or re-execution of unsatisfactory work;
- Suspension of the works;
- Awarding extensions of time;
- Instructions to change the work requirements;
- Measurement and valuation of the works;
- Assessment and valuation of claims;
- Certification of payment and agreement of the final account;
- Resolution of disputes.

Figure 5.3: The Powers of the Engineer in the ICE 6th.

Inevitably there are problems with this approach, particularly concerning the buildability of the design which should, but sometimes does not, consider how the 'innovation' can be physically constructed. Difficulties also occur when the Engineer interferes with the Contractor's preferred method of construction, which should be *"...of a kind and conducted in a manner acceptable to the Engineer."* (Clause 13(2)). Although the Engineer may have good reason to interfere (such as on matters of safety, or concern for the integrity of the permanent works), this is likely to be a source of claim for additional payment.

Initially this interference may occur in the form of an enquiry requesting a 'Method Statement' of the proposed manner of construction (clause 14(6)). This can lay a trap for the Employer in that they can easily become incorporated as part of the stated requirements of the works. This is either done via the requirement for the Engineer's acceptance of the Method Statement (clause 14(7)) or, prior to the award of contract, if received as part of the tender submission[2]. Thus any deviance from the Method Statement (either by Engineer's instruction, or for any previously unforeseen condition) may lead to a substantial claim for additional payment.

Thus the Employer will have to pay for any interference with the Contractor's method of constructing the works which, when considering this competence should be in the charge of the Contractor anyway, is an equitable approach to the responsibility of the works implementation. The question therefore arises whether, for the aforementioned reasons of buildability, the contractor should be given the responsibility for the design of the works on similar grounds of competence. This argument, in favour of Design–Build packages, is considered in Chapter 6.

Maybe it was on the grounds of this that, for the first time in the history of the ICE Conditions of Contract, the 6th Edition allows the contractor to be given design responsibility for part of the works (clauses 7(6) and 8(2)). The contract is silent as to the extent of design. By "part of the works" it is clear that individual work items or even sections of the works could be included, but not the whole of the works.

Following the introduction of the Construction (Design and Management) Regulations 1994 on 31 March 1995, the ICE's drafting committee (the CCSJC) published additional clauses to cater for the provisions of the legislation. However herein lies further potential problems. The new clauses give the Planning Supervisor and Principal Contractor powers to take action under the regulations in regard to Health and Safety matters (Clause 71(3)(a)). The effect is deemed to be an Engineer's Instruction in accordance with clause 13(3) of the contract. Although this may be all very well and necessary for the management of health and safety, this raises potentially alarming issues for the administration of the contract. Although it can be assumed that the Principal Contractor is the same as the main contractor engaged under the ICE 6th, these new clauses introduce yet another player to the contract in the Planning Supervisor. The concern here is that the effect of his actions could have significant implications on the rest of the works in terms of time and cost. Although this was clearly not the intention of the regulations, it would appear that the new ICE clauses in effect have created a direct relationship between the Planning Supervisor and the Contractor.

The introduction of these new clauses has highlighted a number of other amendments which are required to keep the ICE 6th contemporary:

- the Corrigenda to the standard form, accommodating provisions for the New Roads and Street Works Act 1991 and other drafting amendments, published in *New Civil Engineer*, 29 July 1993;
- revision of CESMM2 to CESMM3 in clause 57, as already stated;
- revision of the ICE Conciliation Procedure (1988) to the 1994 publication (clause 66(5));
- consideration of the inclusion of an adjudication clause in accordance with the Housing Grants Construction and Regeneration Act 1996 to prevent referral to the Secretary of State's Scheme for Construction.

Subcontracting under the ICE 6th

The passage to statute of the Housing Grants Construction and Regeneration Act 1996 has had a significant impact on the form of sub-contract the Contractor may adopt. At present there is no ICE form of sub-contract, other than the FCEC Form of Sub-Contract (the 'Blue form', published September 1991) which was recognised as the standard form of sub-contract to be used with the ICE 6th. This is a 'back-to-back' contract in that its terms directly reflect the provisions of the ICE 6th.

Whereas this might not be of great concern to the Engineer or the Employer, the nature of some of the clauses in the way they sit with the ICE 6th, is of particular issue with most sub-contractors. The Blue form adopts 'pay-when-paid' clauses which means the sub-contractor only receives his money once the Employer has paid the main contractor. When one considers the combined duration of clause 60 of the ICE 6th and clause 15 of the Blue form is between 42 and 70 days, there is plenty of time for the sub-contractor to become impatient between actually doing the works and getting the money for it. Furthermore, irrespective of whether the sub-contractor has satisfied the terms of his contract with the Contractor, he will only receive payment if the Engineer has certified it in accordance with the provisions of the main contract. These requirements may be different.

Thus if payment is delayed or not granted by the Employer due to some failure of the contractor or some other sub-contractor, the 'innocent' subcontractors will not receive payment for their effort. A similar scenario would prevail in the case of the client becoming insolvent. Thus when Latham called for legislation to remove unfair contract terms, he included 'pay-when-paid' clauses.

This is now history: the effect the Housing Grants Construction and Regeneration Act has been to outlaw the FCEC Blue Form, thus rendering the ICE Conditions of Contract without a representative standard form of sub-contract; especially as the FCEC was wound up in 1996.

Summary of the ICE 6th

This review has expressed a number of concerns with the ICE 6th Edition Conditions of Contract which have the potential of adversely affecting the delivery of client's business objectives. In short, these concern:

- the dominant role of the Engineer who is expected to be impartial for his/her actions and yet has immunity from the contract;
- several of the supplementary contract documents are outdated and/or conflict with new legislation;
- there is no standard form of subcontract for use with this contract.

In this analysis it is clear that the contract takes control from the client by giving considerable power to the Engineer and his staff (which is perhaps hardly surprising for a contract published by the Institution of Civil Engineers). The contract clearly protects and enforces the role of the Engineer as the 'gatekeeper' to the built environment, but the commissioning client should question whether this is really desirable. Where there are 'winners' (the engineering profession) there are also 'losers' and these will include the construction contractors (who are at the mercy of the Engineers' impartiality) and the paying client.

Overall, the risk allocation between the parties is relatively fair. The client is required to accept the risk of the unforeseen (e.g. ground conditions) and any changes to the works, while the Contractor assumes responsibility for his own workmanship and methods. Once again, however, the Engineer escapes lightly; if this is not addressed in the Engineer's terms of engagement the client accepts the risk of the design being adequate and the Engineer remains immune from liability.

In summary the ICE 6th favours the engineers by giving them power with immunity from responsibility, at the client's expense.

FIDIC 4th Edition

FIDIC is an acronym for the *Federation Internationale des Ingenieurs-Conseils*, which is an international civil engineering construction confederation based in Lausanne, Switzerland.

Members include the Confederation of International Contractors Associations, the Associated General Contractors of America, the European International Contractors, the World Bank, the Inter-American Development Bank, the Asian Development Bank and a group of Arab funding agencies. Among other activities, FIDIC is responsible for publishing an 'international' set of Conditions of Contract in English, commonly referred to as the 'red book' (but not to be confused with the IChemE 'Red Book' Conditions of Contract, examined in Chapter 6). This section examines FIDIC's 'red book': the FIDIC 4th Edition Conditions of Contract (published 1987).

Although its intention is to provide a truly international form of contract, FIDIC is not. Care should be taken by the client to ensure its compatibility with the local or governing statutory law. For example, Sawyer and Gillott[3] state there are many parties who prefer to nominate English Law to govern the interpretation of the contract, however it should be noted that the legal system of a country in which a dispute is resolved (by the courts or arbitration) may require the application of its own governing law and may negate any such choice.

Differences in legal systems can have significant effect. Under English Law there is no concept of *force majeure* unless it has been specifically incorporated in the contract terms (as in the JCT contracts). However the French civil code (*Code Napolean*) recognises *force majeure* as an intervening event and, in certain circumstances, allows for frustration of the contract.

Therefore, in an attempt to prevent the contract being too generalised, FIDIC is published in two parts. Part I comprises a set of mandatory 'core' clauses and Part II contains a number of suggested supplementary (or replacement) clauses. Thus the client is given a suite of clauses to choose the most appropriate allocation of risk depending on the circumstances and locality of the works. *The client is therefore expected to prepare an individual document for each contract.* As a result FIDIC contracts tend to be drafted for larger construction schemes, where funding and consortia of construction professionals are often multi-national.

Well-known projects using the FIDIC template include London Underground's Jubilee Line Extension, the Second Severn Crossing and the Greater Cairo Wastewater Project. Similarly some client organisations, such as PowerGen plc, have adopted FIDIC as their standard form for all major projects. PowerGen has adopted an amended form of FIDIC, called the EPC (the Engineering Procurement Contract), for all its major construction works both in the UK and overseas.

Although it is tempting for many clients to re-draft/amend the core FIDIC clauses, this practice is not recommended in any of the established commentaries. It should only be practised in situations where it is deemed absolutely essential, and even then by professional lawyers familiar with the idiosyncrasies of the local governing laws.

The FIDIC Conditions have been drafted along very similar lines to the ICE 6th Edition Conditions of Contract in terms of their structure, operation and, even, some of the clause numbering. Thus users of the ICE 6th will be familiar with the workings of FIDIC, although no comparison between the contracts should be made and users should not be misled in thinking that one resembles the other. FIDIC is not necessarily subject to English Law and some of its wording, definitions and procedures are very different from those in the ICE 6th.

One of the greatest differences is to be seen in the role of the Employer who is far more active in the contract than most other standard forms of contract; but this is to be expected of major international construction projects where finance, and its control, takes a far higher profile generally.

In short, the FIDIC Conditions offer a sequential contracting strategy for civil engineering construction which separates the roles of design and construction and has a professional third party (the 'Engineer') preside over the contract administration in the form of tri-lateral governance. In this respect the contract could be compared with ICE 6th Edition, the GC/Works/1 or even the JCT'80 building contract.

FIDIC is an admeasure ('measure and value') contract using a Bill of Quantities to schedule a full list of tendered rates for each individual work element of the construction. There is no 'contract sum', as the final account is dependent on the re-measured value of the works. Thus at the point of award, as with the ICE 6th, neither party actually knows how much the final cost of the works will be. It should also be noted that the rules of measurement (such as CESMM3 or SMM7) are not stipulated by FIDIC and therefore need to be specified in the tender documentation.

As with the ICE 6th, the FIDIC 4th Edition allows for some design responsibility to be given to the contractor. Liability is expressed as that of 'reasonable skill care and diligence' (clause 8.1) and not a fitness for purpose requirement. The contract is not a design-build form, but the presence of this feature allows the client some flexibility in choosing who should be responsible for different parts of the design (based on the competencies of the vendor).

FIDIC 4th clauses requiring the Employer to be consulted, along with the Contractor, by the Engineer.	
6.4	Delays and cost of delay of drawings.
12.2	Adverse physical obstructions or conditions.
27.1	Delay and cost of removing fossils, etc.
30.3	Transport of materials or plant – damage to bridge or road.
30.4	Waterborne traffic – damage to structures.
36.5	Engineer's determination where tests not provided for.
37.4	Rejection (repetition of tests).
38.2	Uncovering and making openings.
39.2	Default of contractor (removal of improper work).
40.2	Engineer's determination following suspension.
42.2	Failure to give possession.
44.1	Extension of time for completion.
46.1	Rate of progress.
49.4	Contractor's failure to carry out instructions (remedying defects).
50.1	Contractor to search.
52.1	Valuation of variations.
52.2	Power of engineer to fix rates.
52.3	Variations exceeding 15 per cent.
53.5	Payment of claims.
64.1	Urgent remedial work.
65.5	Increased costs arising from special risks.
65.8	Payment if contract terminated.
69.4	Contractor's entitlement to suspend work.
70.2	Subsequent legislation.

Figure 5.4: FIDIC Clauses Requiring Action from the Client.
Source: Potts[4].

As previously noted, FIDIC is administered on the basis of tri-lateral governance using an 'impartial' Engineer (clause 2.6) in an agency role acting for the Employer. However unlike other standard forms of sequential contract, the Employer has a far more significant role than the usual 'hands-off' approach. Sawyer and Gillott state: *"...There are 23 occasions when the Engineer must consult the Employer and the Contractor and only one when the Employer consults the Contractor on a matter of settlement. The Employer is the more important Party to the Contract because it is his Project and his money which provides employment for the Engineer and the Contractor"*[5]. Figure 5.4 details the clauses under which the Employer is to be consulted by the Engineer.

However good this may appear for the client who wishes to retain a hand of control in the contract administration, there are some limitations with this approach. For example: what does 'consultation' really mean and to what level of detail should it take place? (Presumably it is something between the extremes of a 'notification' process and a 'submission for acceptance' process). Once consulted, how much input does the client have? Furthermore (and equally importantly), once consulted what liability is placed on the client to respond? Some of these questions are tackled individually in the respective clauses, but nevertheless there is a lack of clarity within the drafting.

Other distinctive features in the FIDIC Conditions include:

- **Bonus for early completion** – this is an optional clause in Part II (clause 47.3) which the client may elect to employ. The bonus takes the form of a sum (specified at pre-tender stage) for each day that is saved prior to the completion date (i.e. a reverse form of liquidated damages).

- **Limit on Variations** – clause 52.3 states that if upon take-over the sum of variations (additions or deductions) is in excess of 15% of the "Effective Contract Price" (i.e. the accepted tender total less dayworks and provisional sums), then the contract price should be adjusted after consultation. However there are problems in the wording of this clause. For example: on a £100m major project, if the additions total £25m and the deductions total £20m, then does this mean the variations are

£45m or a net £5m? In the former case the variations exceed 15% by £30m and it is not clear whether the adjustment should be made for the total £45m or just the £30m excess. In the latter case, the variations do not exceed 15% and no adjustment is allowed, however warranted the contractor feels his case may be!

- **Financial Considerations** – as could be expected of a contract for large internationally funded projects, FIDIC has full financial management considerations including: retention of monies (clause 60.2(a)), interest penalties on overdue payments (clause 60.10), fluctuations for changing labour/materials costs and inflation (clause 70.1) and an allowance for other currency exchanges (clauses 71.1 and 72.1).

- **Dispute Resolution** – somewhat surprisingly, the FIDIC Conditions have selected the traditional civil engineering methods of dispute resolution: an Engineer's Decision (clause 67.1) followed by Arbitration (clause 67.3) using the 'Rules of Conciliation and Arbitration of the International Chamber of Commerce'. There is an opportunity for an interim 'amicable settlement' attempt (clause 67.2), but no reference is made to adjudication or other quick and popular ADR methods. In the light of the Housing Grants Construction and Regeneration Act 1996, this could be a significant oversight for those wishing to contract under English Law.

Summary of FIDIC 4th

To sum up, the FIDIC Conditions provide a versatile template for a civil engineering construction contract in the international environment. The contract advocates a sequential contracting strategy with partial design allowance for the contractor. There are considerable dangers in over-looking the governing statutory laws which may apply or, even, supervene the conditions expressed in FIDIC, but this has been partly addressed in providing a suite of optional supplementary clauses in Part II of the contract. Obvious concerns include:

- FIDIC's compatibility with the local governing law;

- the high costs of transaction required to draft individual contract documentation for each works project;
- the immunity of the Engineer from his/her actions;
- the meaning and extent of the 'consultation' between the Engineer and the client.

As with the summary of the ICE 6th, the Engineers are the clear winners in this contract as considerable power is placed in their hands. However this is regulated more effectively than the ICE 6th by allowing the client a greater contribution in the works proceedings. The obvious corollary of this is to recognise that the contractor is marginalised in these proceedings: the power-base is shared by the client who drafts each specific contract and an 'impartial' Engineer on their pay-roll.

JCT Standard Form of Building Contract 1980

The JCT'80 replaced the much criticised JCT'63 which was described by one commentator as overlong, complex and obscure[6]. Despite its poor reputation in the industry, its reputed defects and some of its perceivably difficult procedures, the JCT'80 is still in surprisingly wide use. As the Standard Form of Building Contract, it provides the building block for other JCT contracts. It represents the sequential approach to procuring building works in that it separates out the design and construction elements of the works and has a professional third party preside over the contract administration (the architect).

At the point of award of contract under the JCT'80, the Employer is agreeing to the Contractor's tender to construct the works in accordance with a prescribed design. The design will have been carried out under separate agreement (usually by the architect to the contract), thus the JCT'80 contract is only for the construction element.

Types of JCT'80 Contract

There are six variants of the JCT'80 to reflect public/private clients and different types of reimbursement, as illustrated in Figure 5.5, overleaf.

Despite these various permutations, the differences are subtle and not of significant substance to warrant individual examination in this text. In essence they comprise:

Private/Local Authority

The main difference is in the role of the architect, which for the local authority forms is called the architect/contract administrator (or sometimes the 'Supervising Officer'). The rationale behind this is that in some local authorities, the designated employee to carry out this function may not be a qualified architect. As time has passed and the JCT'80 has been revised by published Amendment, the difference between Private and Local Authority forms has become significantly reduced.

The JCT'80 Standard Forms of Building Contract:			
	Type of Reimbursement:		
Type of Client:	Lump Sum	Approx Quantities	Admeasurement
Public	1	3	5
Private	2	4	6

KEY:
1.	=	JCT'80 Local Authority Without Quantities;
2.	=	JCT'80 Private Without Quantities;
3.	=	JCT'80 Local Authority With Approximate Quantities;
4.	=	JCT'80 Private With Approximate Quantities;
5.	=	JCT'80 Local Authority With Quantities;
6.	=	JCT'80 Private With Quantities.

Figure 5.5: Six variants of the JCT'80.

With Quantities

The contract is reimbursed on a 'measure and value' basis whereby the Contractor tenders a schedule of rates (a Bill of Quantities based on the SMM7 measurement rules). As the works are carried out, the quantities are re-measured and the contract sum is adjusted with the variations in measurement.

Approximate Quantities

The origins of the 'With Approximate Quantities' form date back to October 1975, these were subsequently revised in the JCT'80 form, although their use is becoming less popular today (In 1981, 14.2% of sales of the JCT'80 were 'With Approximate Quantities' forms, but in 1989 they constituted only 7.8% of JCT'80 sales[7]). The contractual provisions are very similar to the 'With Quantities' form except that the quantities stated in the Bill of Quantities (and thus the tender) are only approximate. The works are measured as they are executed, but the quantities are only varied if the original approximation was not an 'accurate forecast'. Thus the final account (in clause 30.6) is built up from 'nothing' as the work is done with any additional claims accrued.

Without Quantities

This is a lump sum contract, tendered on the sum of a build-up of sums from a Bill of Quantities basis (again using SMM7). As such the contractor undertakes to carry out the work described and quantified in the Bill for a stated sum of money. This is not subject to admeasurement and thus the risk of inaccurate quantities is transferred to the contractor. With this form the tenderers will want to be satisfied that the architect's design has been fully and adequately quantified in the Bill of Quantities, as there will be no recourse for re-measurement once the contract has been awarded. The contract sum only changes as the account is varied due to variations, price fluctuations and claims events, *inter alia*.

General Provisions of JCT'80

Thus, although there may be some limited certainty in the price of the works to the client in the 'Without Quantities' form, nevertheless for each type of JCT'80 contract *the full cost of the works to the Employer is not known until the Contractor has completed them.* Furthermore, recent empirical evidence has indicated that on average 20% of the final account sum is due to claims negotiated long after the contract has been signed[8].

Whereas with the re-measured forms of JCT'80 the Contractor has the assurance that he will be reimbursed for his endeavours *pro rata* (based on his tendered rates), with the lump sum versions the Contractor does not have this certainty. Thus in attempting to create certainty for the client, the contractor is required to bear the risk of inaccurate quantities.

This could be considered unreasonable, in view of the fact that the client's advisers have prepared the design, its specification and the Bill of Quantities in advance. As it is the client's design, it follows that the client should be liable for the adequacy of the quantities in the invitation to tender. Although theoretically the client will be paying for this increased transfer of risk to the contractor in the form of a premium, in the current highly competitive market most tenderers will be reluctant to add such a premium if it runs the risk of losing the contract. This could then lead to strained relations during the progress of the works, especially if the quantities were found to be inaccurate. Therefore in the absence of absolute certainty with the quantities, a lump sum traditional contract (such as the JCT'80 "Without Quantities") can only be assumed to be an opportunistic attempt by the client to get 'something for nothing'.

Once the works have been measured and valued by the Quantity Surveyor (if appropriate), the Architect certifies payment for the amount due (clause 30.1.1.1). The Employer is given 14 days to pay the Contractor and has only limited entitlement to pay any less (as discussed below). However despite the fact that Contractors are currently dependent on cash flow to survive, the JCT'80 gives no provision for the recovery of overdue payments or the interest that accrues on them. The Contractor's only recourse is to sue on the certificate asking the courts for a summary judgement under R.S.C. Order 14. This is a relatively quick procedure and in most cases the contractor will receive payment and recoup the costs of action. Alternatively in the event of non-payment the Contractor has the right, under clause 28.2.1.1, to determine the contract. Either action is a drastic remedy that should be seen as a last resort, and it is certainly not conducive to good working relations between the parties. Elsewhere, it has been suggested that it would be better to allow a provision in the contract for the Contractor to temporarily suspend work until payment is received[9].

Under the JCT'80, the client has the right to make limited deductions from the amount certified for payment. Once, an Architect's Certificate was 'as good as cash', but now the client can exercise the right of 'set-off' as well as other deductions for retention (clause 30.4) or for liquidated damages (clause 24). The right of set-off has always been a contentious issue. Essentially the client may withhold payment that is validly due for work done to account for outstanding monies owed for other parts of the works (i.e. money owed is set-off against money due). However this is strictly limited to the bounds of the contract such that set-off from other works contracts cannot be exercised. Under clause 30.1.1.3, the Contractor must be informed of the reasons for any deduction in payment, which gives him the opportunity to contest the decision.

It should be noted that the nature of building works is different to that of civil engineering works and this is reflected in the JCT'80. Building works carry a greater level of certainty in that the extent of work required can be foreseen and prescribed in greater detail. As a general rule they comprise very little work in the ground or with water, whereas civil engineering infrastructure projects do. The uncertain nature of the ground has a significant impact on the works content and therefore a far greater risk of cost/time implications on the overall works. Thus works with fewer elements of ground works will be less uncertain and easier to prescribe. In fact unlike the ICE 6th, the JCT does not have an 'unforeseen ground conditions' clause *per se*. Nevertheless it offers recovery of loss to the contractor for such matters where there is discrepancy or divergence between the contract drawings/documents and the work done (clauses 25/26).

Other factors reflecting the JCT'80's suitability for building works include:

- the wide use of nominated sub-contractors;
- the role of the architect;
- the insurance provisions.

Nomination under the JCT'80

Nomination occurs where the Client's agent (Architect/Engineer) selects the sub-contractor/supplier to be used and instructs the contractor to place a subcontract with them. This will be done either as part of the tendering process (the nomination having been 'named' in the invitation to tender) or in the course of the works (catered for by a prime cost item in the Bill of Quantities). Either way, herein lie many contractual difficulties regarding issues of responsibility, liability and reimbursement; particularly in times of default. Despite the problems, nomination remains a popular means of controlling the supply of specific goods (products/materials) and services to the works. The practice occurs in both civil engineering and building works, but more so in building as a consequence of the architect's professional selection of specific products. Thus the JCT'80 has devoted Part 2 of the contract to nomination and the JCT has prepared specially written practice notes, sub-contracts and other documentation (such as the '1991 Procedure') for it. Nevertheless the practice remains problematic (usually for the client) and is an issue which is probably best avoided if possible. This can be done through greater use of descriptive performance-based specifications, rather than creating problems by naming specific brands of products.

The Role of the Architect

The administrative powers of the architect are geared to the supervision of materials and workmanship and less towards the contract administration. Thus all materials and goods (clause 8.1.1) and workmanship (clause 8.1.2) are to be to the satisfaction of the architect, who is given wide powers to reject and re-execute/substitute works until it meets this 'standard'.

Relatedly the architect's supervisory powers of administrative matters, such as the control of progress, are particularly weak (Figure 5.6 lists their powers under the contract). Under clause 5.3, the Contractor is to give a master programme to the architect. There is nothing in the contract describing what information the programme should contain. Thus a simple list showing the order in which the works shall be completed will suffice; there is no requirement to give any indication of timescale or the resources to be employed. Furthermore in the JCT'80 there is no mechanism by which the architect may approve or reject the programme, or make amendments to it. Similarly there is no means (other than by threat of determination) by which the architect can instruct the contractor to expedite progress. The only requirement on the contractor is to proceed 'regularly and diligently' with the works (clause 23.1), whatever that means!

- Consent to sub-letting the Contract;
- Correction of ambiguities, issue of drawings and revision of information;
- Approval of materials;
- Removal of any personnel from the site;
- Instructions for exploratory investigations;
- Instructions for tests and/or investigations on materials and workmanship;
- Removal and/or re-execution of unsatisfactory work;
- Direction of other supervisory staff (the Clerk of Works and/or QS);
- Postponement of the works;
- Awarding extensions of time;
- Instructions to change the work requirements;
- Assessment and valuation of claims;
- Certification of payment and agreement of the final account.

Figure 5.6: The Powers of the Architect in the JCT'80.

Insurance provisions under the JCT'80

Building Insurance is a specialised field which is beyond the scope of this text[10], nevertheless the JCT'80 provides detailed insurance provisions in clauses 20, 21 and 22 that are particular to the building industry. These are:

- **Indemnities:** injury to persons and property (based on culpable negligence or breach of statutory duty);
- **Insurance:** injury to persons and property (third parties);
- **Insurance:** liability of the employer (clause 21.1) with the exception of the 'excepted risks' listed in clause 1.3;

- **Insurance of the Works:** either new works or existing structures (to cover either the 'specified perils' listed in clause 1.3, or 'all risks insurance'); refer to clauses 22A, 22B or 22C;
- **Insurance:** liquidated damages cover (clause 22D);
- **Remedies if a party fails to insure:** (clause 22).

These provisions are considerably more detailed than those of the ICE 6th, perhaps further highlighting the differences between building and civil engineering works. The additional insurances are to cover liability for damage arising from the works to other property(ies) , *inter alia.*

Other issues

Finally, it is worth noting that:

- the JCT'80 does not provide for any contractor's design in itself. Thus under JCT contracts the contractor's design element is either full (as in the JCT'81) or nothing. However there is now a supplement to be used in conjunction with the JCT'80 which allows the contractor to design a portion of the works[11]. Nevertheless, care should be exercised with this, as the contractor has no express liability for the design save only to inform the client of suspected defects;
- the contract contains many phrases which are open to subjective interpretation (and therefore they are open to abuse if interpreted with opportunism). Examples include: *"...to the architect's satisfaction...", "...in the opinion of the architect...", "...so far as procurable...", "All work shall be carried out in a proper and workmanlike manner",* etc. The problem with these clauses is that the lack of precision allows differences of opinion and conflicts of interests to be fostered. This can escalate into full-blown conflict and dispute in the right conditions, as discussed in Chapter 12.

Summary of the JCT'80

This section has examined the JCT'80 and highlighted some particular issues of concern. It has not been written as a legal commentary or an exposition of the contract (for this the reader's attention is drawn to the relevant published guides listed at the end of this chapter). The main areas of concern have included:

- the role of the Architect, liability for his actions and his inability to control matters of progress;
- the 'all or nothing' Contractor's design element;
- the lack of recourse available to the Contractor in the event of non-payment;
- the problems of nomination;
- the subjectivity inherent in interpreting the contract terms;
- the risk of inaccurate quantities in a 'Without Quantities' lump sum contract.

Like the ICE 6th, the JCT'80 places considerable power and authority in the hands of the Client's agents, the Architect and QS. Again this infers that the Contractor and its suppliers are disadvantaged and subject to decisions which, at times, are beyond their control.

Despite being placed in a position that is arms-length to all decisions, contractors are expected to 'trust' that they will be paid what is justly theirs on time. Such a risk undertaking costs clients a premium and, again, it should be noted that the only true winners are the professional parties whose institutions publish the contractual forms.

Partial Design Responsibilities in Sequential Contracts

There is a somewhat philosophical debate that responsibility for design cannot be absolute; even in the Design and Build forms of contract the Employer must prescribe something of his requirements (i.e. the 'genesis' of design) and conversely in the most prescriptive of traditional contracts the contractor is still required to choose which materials should meet the requirements of the specification. Design responsibility should not be apportioned with strict liability; it must be shared *appropriately* between the parties.

Notwithstanding this, some sequential contracts require the Contractor to take responsibility for both the design and the construction of part of the works and, as previously mentioned, the liability for this is usually limited to that of 'reasonable skill care and diligence'.

This gives the proactive client an opportunity to consider in advance the appropriate competencies required to implement the works and to apportion responsibility to those best placed to manage it. Thus with the ICE 6th, FIDIC, JCT'80 and the NEC, there are options which allow the client to employ the best design competencies from the supply-base. Some design work can only be completed after works have commenced (e.g. in the opening up of a complex excavation) and these provisions are ideally suited to these circumstances.

There is one small word of caution, however, for the client who loses control of the design process and allows the main designer to pass incomplete designs to the contractor. The contractor is then required to submit a design to the Engineer/Architect for consent, which assumes that the agent has the competence to check and approve it. This has potential ramifications on the works programme in the event of delays (the client has lost control of the process). There is also the argument that, if the engineer or architect is capable of checking and approving the design, then why should they not complete it in the first place?

Sequential Contracts: General Conclusions

This chapter has demonstrated several key learning points concerning the balance of power in sequential contracts. Although marked differences exist between the ICE 6th, the FIDIC 4th and the JCT'80 and despite having been written for different markets within the industry, each of the contracts share similar principles concerning the allocations of risk and power. Underscoring each of these analyses has been the protectionism of the industry's professional institutions. These contracts seek to separate power and authority from risk and liability. This is achieved through the principle of *tri-lateral governance*, where a professional third party is paid to administer the contract without recourse for their actions. The ICE 6th favours the engineers, while the JCT'80 favours the architects and quantity surveyors. Only in the FIDIC

4th Edition is there any attempt to regulate the power of the engineer, albeit at the expense of the contracted suppliers. If clients *must* use these contracts, then they need to ensure they engage their third party professionals on terms which match their responsibilities.

Further Reading

There are many good texts and commentaries which cover the provisions of these specific contracts. As suggested previously, these tend to concentrate on the legal and technical details. The following lists are some recommended texts which the authors found helpful in reviewing these forms of contract:

The ICE 6th

- Eggleston B. (1993) *The ICE Conditions of Contract 6th Edition: A User's Guide*, Blackwell Scientific Publications, Oxford.

FIDIC 4th Edition

- Bunni, G. (1991) *The FIDIC Form of Contract: The Fourth Edition of the Red Book*, BSP Professional Books, Oxford.
- Jones, G. P. (1979) *A New Approach to the International Civil Engineering Contract: A detailed analysis of the FIDIC International Form of Civil Engineering Contract*, The Construction Press Ltd, Longman, Lancaster.
- FIDIC (1989) *Guide to the use of FIDIC Conditions of Contract for Works of Civil Engineering Construction*, 4th Edition, FIDIC, Lausanne.
- Sawyer J.G. & C.A. Gillott (1990) *The FIDIC Digest: Contractual Relationships, Responsibilities and Claims under the fourth edition of the FIDIC Conditions*, Thomas Telford, London.

JCT'80

- Chappell, D. (1995), *Understanding JCT Standard Building Contracts, 4th Edition*, E&FN Spon, London.

- Fellows, R. F. (1995), *1980 JCT Standard Form of Building Contract: a commentary for students and practitioners, 3rd Edition*, Macmillan Press Ltd, Basingstoke.
- Price, J. (1994), *Sub-Contracting under the JCT Standard Forms of Building Contract*, Macmillan Press Ltd, Basingstoke.

Chapter Notes

1. Cottam G. & G. Hawker (1991) *ICE Conditions of Contract for Minor Works: A User's Guide and Commentary*, Thomas Telford, London.
2. Refer to the case of *Yorkshire Water Authority v. Sir Alfred McAlpine & Son (Northern)* [1985] 32 Build L.R. 114.
3. Sawyer J.G. & C.A. Gillott (1985) *The FIDIC Conditions Digest of Contractual Relationships and Responsibilities, 2nd Edition*, Thomas Telford, London.
4. Potts K. (1995) *Major Construction Works - Contractual & Financial Management*, Longman, Harlow, p. 270.
5. Sawyer J.G. & C.A. Gillott (1990) *The FIDIC Digest: Contractual Relationships, Responsibilities and Claims under the fourth edition of the FIDIC Conditions*, Thomas Telford, London, p. 5.
6. Eggleston B. (1992) *Liquidated Damages and Extensions of Time in Construction Contracts*, Oxford, Blackwell Scientific Publications, p. 206.
7. JCT (1990) *The Use Of Standard Forms Of Building Contract*, RIBA Publications Ltd, London.
8. Farrow J.J. & G.W. Wagland (1991) *Effectiveness of Claims Provisions under the JCT Standard Forms of Contract*, The Chartered Institute of Building, Occasional Paper No. 47, p. 14.
9. Powell-Smith, V. (1993) 'Put Penalties for Non-Payment into Contract' *Contract Journal*, 14 October 1993.
10. For a comprehensive guide, refer to: Wright J. D. (1997) *Construction Insurance: Practice, Law, Reinsurance and Risk Management*, Witherby & Co. Ltd, London.
11. Practice Note CD/2 Contractor's Designed Portion Supplement.

Chapter 6

Design & Build Contracts

Introduction

The last two decades have seen a marked increase in the use of design and build (D&B) contracts. The move was spear-headed by the publication of the JCT'81 Standard Form of Building Contract with Contractor's Design written for the building industry at large in 1981. It was published in response to requests from the Department of the Environment to represent those clients wishing to obtain both the design and consequential construction from the same supplier. Although this was not the first D&B contract (the first edition of the IChemE Red Book was published in 1968), it was seen as the first standard form of D&B contract that was wholly appropriate to the construction industry.

The 1980s decade saw D&B contracting develop from being a novelty to as much as 40% of all new building work procured this way. Towards the end of the 1990s this figure has declined, but still stands at an impressive 20% of all construction contracts awarded.

D&B contracting has received much public acclaim as well as some criticisms too. This chapter explores the power structure inherent in the approach, as well as the perceived strengths and weaknesses of D&B contracting and three of its most common standard forms of contract:

- the ICE Design and Construct contract;
- the JCT'81; and
- the IChemE Red Book.

Like the other chapters in this Part, this examination is not written as a legal commentary or as an exposition of the finer details of the contracts' text; for this the reader is guided to the list of further reading towards the end of the chapter. The purpose of this chapter is to provide the reader with an overview of the general business principles behind this type of contracting and its common contracts.

What is Design and Build?

Much discussion has been given to the D&B concept, despite its simplicity. The previous chapter considered the popularity of sequential contracting, where the design is hived-off from the construction elements of the works and contracted separately with professionals. This concept was mainly initiated through the exclusivity of professional institutions and other 'clubs', which sought to differentiate a 'professional' element (as in the design and project planning/management) from the trades element of the physical works. Having given these 'professional' elements an institutional status, the members of those professional institutions were able to charge inflated fees for their services to the industry.

Thus, a two-tier approach to the delivery of construction works developed. As time progressed, more professional institutions became recognised with the result of more 'professional' parties being required in the construction process. Consequently the process became increasingly fragmented as clients were required to find an increasing number of suppliers and professional advisers to help deliver the finished works to requirement. Much of today's problems concerning the apparent necessity to 'construct the team' would never have occurred if the industry was not as fragmented as it is.

With this structural fragmentation comes a fragmentation of interests. Opportunism is increasingly manifest as each of the parties becomes more and more self-interested in the fight for command of the ever-scarce resources and fluctuating workload. Thus, Latham's notion that the industry should shrug off its adversarial nature and attempt to generate a team environment by trusting one another is a noble, albeit naive, idea.

This discussion has been elaborated by Cox and Townsend[1] who argue that a focus on structural appropriateness, based on relational competence, is the starting place for any firm attempting to maximise its business objectives.

Notwithstanding this argument (which this publication supports and complements in its discussions), the design and build contracting method cuts across the fragmentation created by professional institutions and their associated 'clubs'.

The design and build approach is based on the principle of *single point responsibility*. This is discussed more specifically in the following section. Quite simply it means that instead of trying to pull together a team of construction professionals with divergent interests, the client forms a contract with one single entity (a main contractor) who is given the full responsibility of delivering the completed works to time and requirement (or in plainer language: one person takes the responsibility of banging the industry's heads together). Naturally this incurs costs as well as benefits.

Thus in D&B contracting, the main contractor conducts both the design works and the means of construction to supply the works to the client's prescribed requirement. There is minimal fragmentation, minimal diverging interest, minimal inefficiencies at the interface and minimal costs of transaction. The construction supply chain is more integrated and, assuming the contractor can manage it properly, it is easier to control.

The Benefits of D&B

The perceived benefits of D&B contracting have been broadly disseminated. They include:

- **shorter lead times** between inception and mobilisation onto site, thus enabling a quicker speed to market (the construction can start before the design has been substantially completed);
- **single point responsibility** for the design and subsequent construction has reduced the number of supplier relationship interfaces (and inefficiencies) for the client to manage;
- **fewer claims and disputes** ensue, as the main contractor is responsible for the whole of the works;
- **greater price certainty** as a result of fewer claims and changes;
- **greater assurance of buildability** as the constructor is also responsible for the design;
- **increased likelihood of the most economically advantageous works**, as the contractor tenders on his own capabilities and competencies rather than to someone-else's design criteria;
- **greater potential choice of end-product**, based on the main contractor's ability to use his own competencies and introduce innovative solutions to supply the client's construction requirements; and
- **more risk passed on** to the contractor to deliver the works to time, budget and requirement. This, of course, also carries its drawbacks; the client will pay a price premium for being risk averse. Some research has suggested that tender prices will vary by +/– 5% of the total tendered sum based on the risk allocation of the conditions of contract alone[2].

These points have been supported by empirical evidence[3] from a survey of 332 projects conducted in 1995/96, which claimed:

- D&B contracting is 12% faster than the sequential contracting method and takes 30% less time to deliver a project from the start of design to the works completion;
- D&B contracting is 13% cheaper than sequential contracting methods;
- D&B projects are 50% more likely to finish on time.

This is impressive evidence which would seem to support the wide-scale adoption of D&B contracting. However the same research also observed issues of poor quality associated with D&B

projects, particularly where 'novated' design occurs. This has been seen as a significant problem and drawback of the D&B approach.

It stands to reason that, where one supplier is asked to produce a design specification and then build works to that specification, the supplier will attempt to do so in the most commercially-efficient manner. That is not to say contractors are attempting to cut corners and deny clients construction of fully merchantable quality, it is simply part of the nature of the contract that is formed. Contractors are being asked to tender to the lowest price in order to provide the most economically advantageous works while their average profitable margins vary between 1% and 2% of annual turnover[4]. In today's overtly competitive market structure, it follows that clients cannot expect to get *something for nothing* and that the 'lowest-priced tender' will only win the 'lowest compliant quality'.

Thus clients who wish to ensure a specific level of quality in an element of the construction works, will be well-advised to consider carefully the level of prescription that is placed on tender enquiries and, subsequently, the level of detail specified in the returned tender documents. This point is picked up and elaborated in the following discussions on the common forms of D&B contract.

The ICE Design and Construct Contract

The ICE Design and Construct Conditions of Contract is based heavily on the standard ICE Conditions of Contract format. This is typified in the ICE 6th Edition which was discussed in the preceding chapter. Many of the clauses are similar, albeit with different underlying principles. Whilst the benefits of D&B contracting have been directly realised in the building industry, the civil engineering market has been slower to accept the design–build principle. This has traditionally been due to the uncertainty involved in civil engineering works; particularly where schemes have required considerable work in the ground or in the proximity of water. The absence of a design–build civil engineering contract was also a major restraining factor which, with the publication of the ICE Design and Construct Conditions of Contract in October 1992 by Thomas Telford Services Ltd, has now been overcome.

The contract's Guidance Notes state:

"The ICE Design and Construct Conditions of Contract depart radically from this [traditional] *concept with the Contractor responsible for all aspects of design and construction, including any design provided by or on behalf of the Employer. The Form of Tender provides for payment on a lump sum basis, but other forms* [of reimbursement] *may be used."*

However, in effect, the arrangements for the Design and Construct contract are much the same as the ICE 6th albeit tempered by the lump sum concept. Although many clauses have been lifted directly from the ICE 6th, or only amended slightly, no comparison should be made between the contracts, as the legal interpretation will be taken from the Design and Construct contract itself and not from any other documents, however similar they may appear to be.

The ICE Design and Construct contract is based on the **Single Point Responsibility** principle. That is, it is the Contractor's obligation under the contract to *"...design construct and complete the Works..."* (clause 8(1)). The underlying principle has major ramifications for most of the contractual mechanisms and there is a major shift in the balance of risks between Employer and Contractor. In effect, beyond the Employer's stated requirements the Contractor subsequently has total responsibility to deliver the works *per se*.

As one commentator has stated, the foundation for success in D&B contracting rests with the quality and clarity of the employer's requirements[5]. This is a fundamentally important point. The Contractor can only design, construct and complete what he is told to do; in the event of ambiguity, the *contra proferentum* rule[6] will prevail in his favour. Furthermore, under clause 51(1), the Employer's Representative only has the power to vary the original [pre-tender] Employer's Requirements and not the Contractor's design or interpretation of the requirements.

Thus the Employer's Requirements need to comprise a precise definition of what is wanted in the works. The level of detail is not specified in the conditions of contract: this needs to match the level of detail of the Employer's interest; the Contractor can only 'mirror' the requirements asked of him. Thus, if the design of a handrailing, for example, is of no consequence or interest to an Employer this should not be detailed in the Employer's Requirements. Conversely, if the painted colour or metallic finish of the handrailing is of importance to the Employer, this will need to be prescribed in the pre-tender documentation. The Contractor cannot be expected to have a prophetic gift in interpreting an Employer's Requirements.

The extent of liability for the design only goes as far as *"reasonable skill care and diligence"* (clause 8(2)(a)), as in all ICE contracts. Therefore, even if the full 'intent' of the works is stipulated in the Employer's Requirements, a *fitness-for-purpose* requirement cannot be levied without legal amendment of the terms of the contract.

There are other noteworthy points concerning the Contractor's design responsibility:

- There is no specific requirement on the Contractor to carry insurance for the design, or any other form of professional indemnity. If the client wishes to have this safeguard (and, moreover, to an appropriate level for the works), this will need detailing up-front in the Employer's Requirements.
- Unlike any other standard form of contract, the ICE Design and Construct Conditions of Contract requires the Contractor to institute a quality assurance system (clause 8(3)). Although the specific nature of the system, whether to ISO 9000 or not, is not prescribed. Again, if the client wishes the system to be of a minimum required standard, this will need detailing up-front in the Employer's Requirements.
- Responsibility for the unforeseen still lies with the Employer (clause 12) and the Contractor has the potential to place 'caveats' on all, or any part of, the design on the basis of the information available within the Employer's Requirements. As an aside, some construction clients have introduced their own means of re-allocating this risk. For example, on occasions the Department of Transport has invited two separate tenders, one accepting the risk of the unforeseen and the other where this is

passed on to the contractor. Other clients have amended their standard forms such that suppliers accept the risk of unforeseen conditions.

- Handover to the Employer can only occur once the Operations and Maintenance Instructions have been submitted (clause 48(3)). The level of detail for these manuals has been stipulated as that sufficient for satisfactory operation, maintenance, dismantling, re-assembly and adjustment of the works. They are not 'as built' drawings and if that or any other details are required they should be prescribed in the Employer's Requirements.

Under clause 6(1) the Contractor is required to request any additional information of the Employer that is required to design or construct the works. However, since the contract price (a lump sum) is based on the pre-tender information supplied to the Contractor, any additional information may constitute a variation under clause 51(1) and the Contractor is likely to be alert to this.

Because the tender is based on a lump sum, the Contractor accepts the risk of any variation in measurement by basing his price on his own estimate of the quantities. As there is no allowance for admeasurement in this contract, this constitutes a major shift in the balance of risks from the ICE 6th. Payment is made to the Contractor at 'interim' periods as prescribed by the contract (clause 60(1)). Since there is no reference to monthly valuations, the interim payments could be made either after set time periods, or by completing specific milestone activities.

Without specific milestones detailed, the valuation of requests for interim payment becomes a difficult task (essentially an estimated percentage of the total lump sum). Similarly variations to the contract are equally difficult to assess. Clause 52 requires the Employer's Representative to value additional works on the basis of what is *"...fair and reasonable..."* (whatever that is).The process comprises a quotation of the work from the Contractor, followed by negotiation with the Contractor if the Employer's Representative does not accept the asking price. Furthermore if there is still disagreement the Employer's Representative is expected to determine the sum for the additional works.

This is a wholly unsatisfactory method of procuring works variations and likely to produce all sorts of conflict. The quotation has no requirement to be competitive and since the Contractor is already mobilised and in possession of the site, he can ask whatever price he wants. Thus if negotiation cannot produce an acceptable sum, the Employer's Representative will direct the price to be paid; which is an equally unsatisfactory resolution

For these reasons, it is advisable for both parties to include a 'contract sum analysis' with the tender. This would detail the activities which built-up the Tenderer's sum and could also provide a schedule of rates for works. Thus the interim payments would be determined in line with the milestones and any variations could be valued on a build-up of rates from the schedule. Although this is relatively simple contract procedure and of fundamental importance to the smooth administration of the contract, the ICE Design and Construct Conditions does not address this issue.

Although the principles of this contract are completely different to the ICE 6th and there is a major change in the role of the parties, the contract is still administered by an Engineer. His role is slightly different to that in the ICE 6th with principal duties comprising:

- programme monitoring;
- inspections to ensure compliance with the requirements;
- design checks (see below);
- certification of interim payments and assessment of the Final Account.

Unlike the ICE 6th, there is no requirement for the mode and manner of construction to be approved by the Employer's Representative, nor is there any provision for nominated sub-contractors or provisional sums, as these requirements are all part of the risks carried by the designer of the works. Although this may reduce the burden of site administration on the Employer's Representative, the requirement to ensure the Contractor's design complies with the Employer's Requirements significantly increases his responsibilities. The Employer's Representative is not required to check and approve the design, but he is required to give consent that it complies with the Employer's Requirements. Thus any dissonance between what the client stated that he wanted and what he actually gets will be the responsibility of the Employer's Representative. Unfortunately there is no recourse on the Employer's Representative in the ICE Design and Construct contract.

As a final comment, the FCEC produced a form of subcontract for use in connection with the ICE Design and Construct form in much the same vein as the FCEC Blue form for the ICE 6th. It is a back-to-back form employing pay-when-paid clauses which, as such, conflicts with the Housing Grants Construction and Regeneration Act 1996. As the FCEC no longer exists, there will be no more industry-standard subcontracts for contractors to use.

Summary of ICE Design and Construct contract

This section has examined the ICE Design and Construct Conditions of Contract and highlighted some of its particular features. Following the pattern set down by the ICE 6th, this contract continues to support the role of a third party acting in tri-lateral governance. In this contract, however, this position of power is severely compromised. The single point of responsibility principle over-rules any need for control to be given to a third party and the main role of the Employer's Representative is to ensure that the contractor provides what was specified in the Employer's Requirements. Thus the critical element of control for the client is the production and quality of the Employer's Requirements; for this determines the success of the ensuing contract.

Thus, risk allocation is established with clarity. The contractor accepts all risks for the design and construction of the works, on the basis of the information supplied. It follows that is in the client's interests to supply that prior information as accurately as possible.

The Standard Form of Building Contract with Contractor's Design (JCT'81)

Like the ICE contracts, the JCT'81 contract is based heavily on its principal contract, the JCT'80. Again, it has several parallel clauses, with as much as 75% of the text being identical[7], albeit with different underlying principles.

One of the key issues facing clients concerns which contract is the most appropriate to be used. The JCT has published guidance (Practice Note 20)[8] to advise practitioners when to use the various forms of JCT contract and suggests that where an architect/ contract administrator has been appointed to prepare the drawings, etc. or where only a portion of the works is to be designed by the contractor, the JCT'81 is not a suitable contract and the JCT'80 is preferable. As the basis for this important procurement issue, this guidance is unhelpful and wholly inadequate. The key question for the client – *who should be carrying out which tasks?* – has not been addressed. It is assumed the answer is already known and therefore determines which form of contract (JCT'80 or JCT'81) is to be employed.

Getting the right specification

Before examining some of the provisions of the JCT'81, it is important to understand how the employer's requirements for the building works are prescribed and how the contractor can satisfy the client in producing the design and completed works. This is as fundamental to the successful administration of the JCT'81, as it is to any D&B contract. It should be noted that in the contractual provisions **the contractor does not have to have his design checked or approved by the employer**. As with all standard forms of contract, the employer's agent (or engineer or architect) may intervene and instruct changes; but this will create claims for additional expense over and above the pre-agreed contractual requirements.

Thus the client is well-advised to convey to the contractor as much information concerning the requirements, intention and performance of the building works as is known or desired. This is carried out at tender stage through the 'Employer's Requirements' schedule. The contractor is required to submit proposals as part of his tender to indicate how the Employer's Requirements will be satisfied. This may include important information about the design, or even a design proposal. It is important to recognise that no contractor is legally bound to tender in the manner laid down in the Employer's Requirements. Since the tender represents an offer of contract, it is important that the client assesses the Contractor's Proposals. This can be expressed in the following formula:

Employer's Requirements + Contractor's Proposals = Complete Definition of the Works

Source: Janssens[9]

There are many items to include in the Employer's Requirements when compiling the invitation to tender. These might include:

- a prescription of the work requirements;
- performance issues;
- specification of certain materials and/or finishes;
- requirements for testing or sampling (clause 8.5);
- QA requirements;
- The Health and Safety Plan (required for CDM Regulations);
- the Environmental Statement (if applicable);
- issues relating to planning or other consents (clause 6);
- issues relating to reimbursement (clauses 30, 12.5, S5, etc.);
- insurance provisions (clauses 21.2.1 and 22D.1);
- site availability (clauses 2.3.1, 7, 23.1);
- issues associated with timing and liquidated damages for delay;
- design requirements for checking/approval (supplementary clause S2);
- operations manuals, etc. (clause 5.5);
- parent company guarantee (and/or performance bond).

Beyond these stated requirements, the JCT'81 gives the contractor total responsibility to deliver the works. Thus it is essential that the client has assessed the tendered Contractor's Proposals to ensure they precisely match that which is desired. ***Once accepted, the*** *client is contractually tied to whatever the contractor has proposed.*

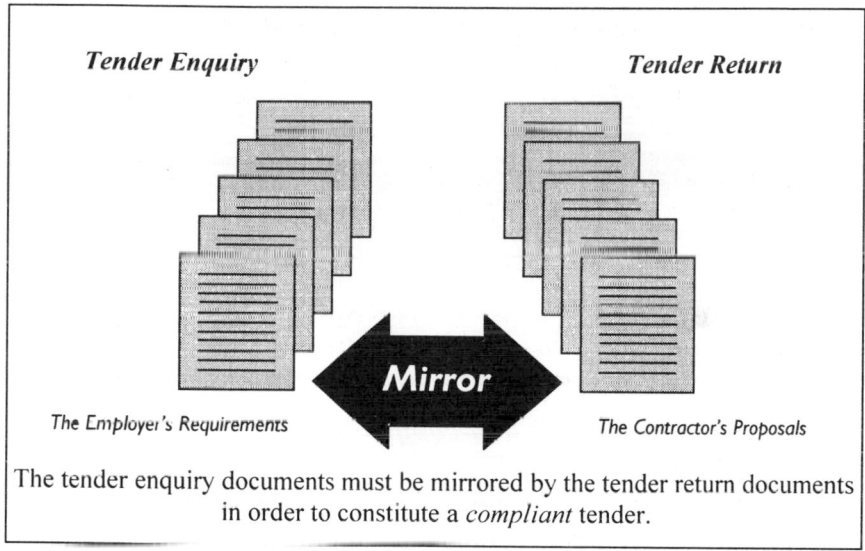

Figure 6.1: A compliant D&B tender.

This may be represented in the illustration in Figure 6.1. If the Contractor's Proposals mirror the Employer's Requirements then the contract can be awarded and satisfactorily fulfilled. However, in practice, the tenderer may wish to qualify certain requirements or even offer alternatives to the client. Furthermore, there may be cases when the Contractor's Proposals fail to match the Employer's Requirements because something is either omitted or incorrectly interpreted. This may be illustrated in the following (overleaf):

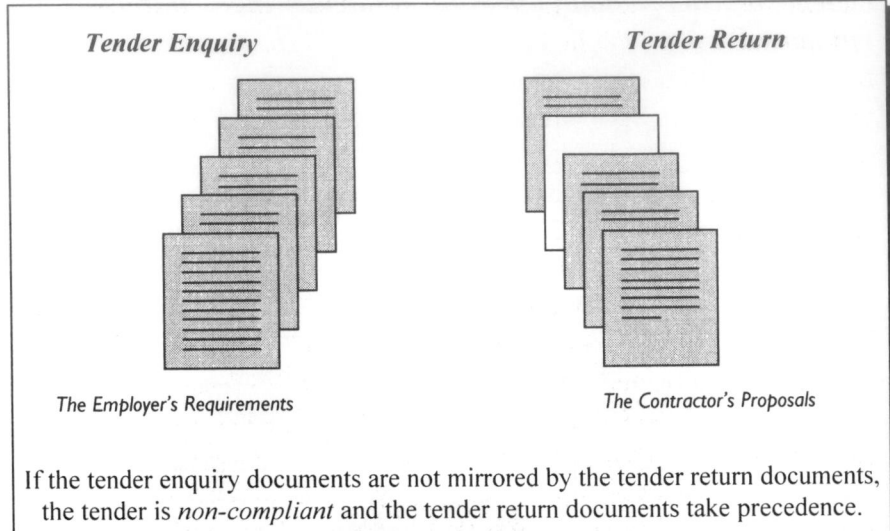

Tender Enquiry

Tender Return

The Employer's Requirements

The Contractor's Proposals

If the tender enquiry documents are not mirrored by the tender return documents, the tender is *non-compliant* and the tender return documents take precedence.

Figure 6.2: A non-compliant D&B tender.

In Figure 6.2, the contractor has not submitted a compliant tender: the tender return documents on the right-hand side do not reflect the original requirements of the tender enquiry documentation. If the tender was accepted the ensuing contract would not deliver the Employer's Requirements (the Contractor's Proposals would supersede the original enquiry) and the contract administration would become problematic; probably resulting in time and cost overruns and/or claims and disputes.

It should be noted that this is essential to the successful administration of the JCT'81, *but not necessarily other D&B contracts*. For example, the ICE Design and Construct contract allows the Employer's Requirements to over-rule any discrepancies in the Contractor's Proposals.

General features of the JCT'81

As a D&B contract, the JCT'81 is based on the *Single Point Responsibility* principle. That is, it is the Contractor's sole responsibility to carry out and complete all works described in the Employer's Requirements, the Contractor's Proposals and the Articles of Agreement (clause 2.1). This is generally the full design of the works and the subsequent construction. The intention is that the works might become design-led; however in practice many contracting firms will sub-let the design to a consultant or architectural practice. Although this may still lead to buildability issues, there is a fundamental shift of risk from the JCT'80 principle, as the JCT'81 expects the Contractor to be fully responsible (clause 18.2.3).

The main issue for clients selecting D&B works is the extent to which the design should be prescribed in the Employer's Requirements. This opportunity allows the intelligent client to consider in advance the appropriate competencies required to design the works (at whatever stage) and to apportion responsibility to those best placed to manage it. Furthermore the client will wish to ensure he has retained adequate control on the actions and performance of suppliers (i.e. the contractor's designers). This, like the ICE contracts, is not addressed in the JCT'81.

Regarding the contractor's liability for the design under the JCT'81, the following points should be noted:

- The contract does not mention the extent of liability on design, other than that the contractor shall have *"...the like liability to the Employer ... as would an architect or ... other appropriate professional designer holding himself out to be competent to take out work for such design..."* (clause 2.5.1). This is clumsy phraseology but, in effect, is no more than the "reasonable, skill and care" liability required by statute (Supply of Goods and Services Act 1982). Thus even if the full 'intent' of the works is expressed in the Employer's Requirements, a *fitness-for-purpose* requirement cannot be levied without amending the terms of JCT'81.
- As with the ICE Design and Construct contract, there is no specific requirement on the contractor to carry insurance for the design or any other form of professional indemnity.
- There is no specific requirement on the contractor to institute a quality assurance system (unlike the ICE Design & Construct form).

- As already mentioned, there is no requirement in the JCT'81 to check or approve the contractor's design; the only requirement is to provide the client with copies of the design details (clause 5.3). However, application of the [optional] supplementary clause S2 allows the client to comment on the contractor's design.

Although single point responsibility prevails, the client is require to employ a supervising officer (the 'Employer's Agent'). There is little within the text of JCT'81 to define the role and extent of the supervision required. Thus the commentary's suggestions as to whom should fulfil this role are divided between the architect and quantity surveyor professions. However common sense would suggest that neither profession is appropriate, as their skills and training have been developed for other tasks. The role and function of the Employer's Agent is fundamentally one of project management, to which a suitably competent project manager should be assigned. Once again, this demonstrates the bias of the professional institutions (the RIBA and the RICS) to protect the self-interest of their members before the industry or their clientele.

One further consideration the client needs to make before contracting under the JCT'81, is the extent to which the Employer's Agent needs to supervise. For instance. there are no programming requirements in the contract. Clause 8.5 requires the client to have stipulated the extent of samples and/or testing required to check the adequacy of the works. If the contractor is to retain single point responsibility, then the client needs to be satisfied with an appropriate level of supervision. Arguably a competent contractor will be self-policing to ensure the works have been built to that which he specified he would in his design. Therefore it follows that the level of supervision should rest on the client's ability to ensure such self-policing/certification is being competently carried out.

Novation

One common source of bad practice is the role of the 'novated' architect. This is an architect commissioned by the client to produce the Employer's Requirements and then subsequently retained (novated) by the contractor to carry out and complete the full design. Some have argued that this transfer of rights allows a potentially smoother flow of design work from one party to the other, as the design does not have to be picked up and developed from another's concept[10]. However this view is too ignorant of commercial reality and the latent opportunism inherent in the 'professional' practices. In practice, the design stages cannot be a continuously smooth process, as they are interrupted by the tendering procedure for the main D&B contract and, furthermore, *any professional* should be able to pick up and develop another *professional's* work 'smoothly'.

Most importantly, the transfer of allegiance under novation is contrary to the client's commercial interests. The novated architect becomes a *single-sourced supplier* in a position of monopolist power over the tendering contractors. This means that, because the main contractor is dependent on the novated architect, both the main contractor and the client suffer the loss of control on cost and quality. The novated architect is able to fix his own rates without regard to the prevailing market because of his monopoly position. Furthermore, novation encourages a degree of 'cosiness' where the sins of omission and/or error remain undisclosed. Recent empirical research has suggested that the worst quality occurs when novation is used[11]. Despite this, it is understood that the Reading Construction Forum uses novation on 50% of all D&B projects.

Reimbursement under JCT'81

Finally, on a point regarding reimbursement, the JCT'81 is primarily a lump sum contract which allows for milestone payments to be detailed in the Contractor's Proposals. A 'Contract Sum Analysis' should be prepared as part of the contractor's tender and thus could be produced in any form. Although bills of quantities are not part of the main provisions of the JCT'81 (see supplementary clause S5), in practice it is beneficial to all parties if the contract sum analysis provides a breakdown of the elements of the works in a form similar to such a bill. Thus in the event of any instructed changes, an appropriate sum of money can be assessed and reimbursed. The JCT'81 contract does not provide for this, so if the client wants to ensure this detail, it needs to be stipulated in the Employer's Requirements at pre-tender stage.

One alternative means of reimbursement to the standard lump sum arrangement in the JCT'81 is the Guaranteed Maximum Price (GMP) system. This is a mechanism which offers the client full price certainty on the contracted works; the contractor accepts the risk of any cost increases. It is an approach which has attracted much discussion and it is not without its critics who suggest that GMP equates to maximum prices guaranteed for the lowest quality works. Notwithstanding this opinion, there are many successful projects being procured through this route.

In Case Study 4, at the end of this section, we examine one GMP project which achieved in excess of 30% cost savings under the management of AMEC Design and Management Limited[12]. The case study demonstrates the utility of the GMP mechanism when it is used appropriately. However it should be noted that this contract amendment was not the only determinant of the project's success; other factors included the adoption of an appropriate collaborative relationship and the main contractor's successful application of value engineering exercises.

Summary of JCT'81

In summary the JCT'81 does not promote strong 'hands-on' supervision or management of the contractor. The Employer's Agent does not have a programme to monitor and there are no requirements to check or approve the contractor's design details. Furthermore, as the contract is based on a lump sum, there is no provision for the breakdown of rates in the event of any changes. These shortcomings should ring alarm bells among some clients.

The balance of power in this contract rests with the ability of the client to express the work requirements in the tender documentation and to ensure that the tendered proposals met the requirements. Any failure in this ability will result in a fundamental switch of power to the contractor; in this scenario the client is powerless to change the works details except through the use of the Employer's Agent and, as a result, being subject to the costs of re-working.

One significant trap for the undiscerning client to be aware of is that of novation. This is the antipathy of best practice, giving power to the architect at the client's expense.

Key questions remaining unaddressed in the JCT'81 include:

- who should be carrying out which tasks? (i.e. how far should the design prescription in the Employer's Requirements be taken and who should be carrying out the task of Employer's Agent?)
- the detail to which the contract sum analysis should be detailed;
- the level of supervision required of the Employer's Agent.

Case Study 4: Peter Black Healthcare Limited[12]

The Peter Black Group manufactures a portfolio of personal care and footwear products. Many of these are 'below the line' products for major blue-chip retailers' own labels. In 1989 Peter Black entered a new market by acquiring English Grains Ltd who have an 18% market share in over-the-counter vitamins and minerals (well-known products such as Red Kooga, Ginseng, Calcia, Natracalm, Natrasleep and Folic Plus). Peter Black was so successful that, within three years, it needed new premises.

In 1993, AMEC Design and Management Limited were contracted to design, manage and construct new facilities in Derbyshire under the JCT'81 contract amended with a GMP mechanism. AMEC undertook to deliver the facilities within the agreed contract sum and at their own risk.

The project comprised 12,879 square metres of laboratories, tableting production, packaging and warehousing facilities as well as offices and other support facilities. The contract sum was set at a maximum of £8.6 million with an additional £2.5 million for Peter Black's process costs.

In the proceeding development stages, AMEC conducted rigorous value management and value engineering exercises to ensure Peter Black got what it wanted and also to challenge them about what they actually *needed*. From these sessions, AMEC were able to re-design and re-specify Peter Black's requirements to ensure they still delivered all that Peter Black wanted, in terms of quality and functionality, but for less overall cost.

In the following construction implementation stages, AMEC adopted teamwork strategies to quicken decision making and encourage early problem solving. Local subcontractors were included in the 'team' and were given clear financial targets for each subcontracted activity aided by revised unambiguous subcontract documents.

One example of the teamwork approach was in the procurement of the metal doors, frames and viewing panels in the laboratory and production facilities. The doors required robust seals without any recesses that might allow bacteria to collect. AMEC approached Henderson-Bostwick who, although well-known for the manufacture of high quality industrial doors, had no previous installation experience of these specific door types. Through teamwork and value management, a new design was produced, based on *functionality*. The doors were manufactured and installed to programme. The total subcontract price was under AMEC's original budget and saved £70,000 from the main contract sum.

The full project was completed within two years of design initiation and three weeks ahead of schedule. The total cost savings were approximately £400,000 and shared equally between client and contractor. Overall the full project cost (at £728/m^2) was 37.9% below the industry benchmark costs (of £1180/m^2) for these type of works.

Although the GMP mechanism on the JCT'81 was only one [small] contributor to the success of the project, it signalled an important principle to the management of the construction process for both client and contractor alike. Without this 'incentive' the contractor would not have accepted the risk of guaranteeing the price, or have the incentive of re-engineering the design specification in order to achieve the highest functionality for the lowest cost.

AMEC have continued to adopt this contractual mechanism in other successful construction projects, as it offers the client a high degree of price certainty, while it means that AMEC can continue to compete on its value management and value engineering competencies.

The IChemE Red Book

This section introduces the provisions of the IChemE Red Book, of which the latest version is the 3rd Edition, published on 31 August 1995. Like the other IChemE Model Form of Contract (the Green Book), the Red Book is a performance-based contract written specifically for construction in the plant and process industries. As such, general points concerning these forms of contract and performance contracts more specifically, are provided in Chapter 9. This section considers the IChemE Red Book and its general application as a D&B form of contract.

The essence of the Red Book is that, within the bounds of the reasonably foreseeable, the extent of works can be 'defined' prior to the award of the contract. This is not the same, however, as the sequential contracts examined in the previous chapter. Under the IChemE Red Book, the contractor undertakes to design, provide and commission the plant to a defined set of performance criteria in return for a fixed lump sum price. The contractor is required to supply all necessary goods, services and competencies required to meet the specification within the timescales set. This specification becomes part of the contract upon award and thus all performance requirements and guarantees must be detailed and known prior to this. Any change of mind by the client, or the introduction of further performance requirements, constitutes a variation and is treated as such, subject to the provisions of clause 16.

Once the contract is in force, the lump sum 'Contract Price' applies, irrespective of the actual costs incurred by the contractor; thus the client can budget for his cost of the works with a fair degree of certainty (notwithstanding changes that occur within the course of the contract). The contractor is reimbursed periodically, as determined by Schedule 8, which may be in the form of monthly or milestone payments. There are provisions for interest in the event of payment becoming overdue (clause 39.5), but there are no provisions for inflation or price fluctuation; these, as with all other cost contingencies, remain at the contractor's risk. Should payment remain outstanding, the contractor may render notice to the client of his intention to suspend works after 28 days. Such suspension remains at the expense of the client who is required to reimburse the contractor his costs plus a 'reasonable' profit (clause 39.6), whatever that is. Furthermore, should the suspension for non-payment last 121 days, the contractor has the right to terminate the contract and retrieve his outstanding monies accordingly (clause 39.7).

The contract is governed by an impartial professional third party: the 'Project Manager' (clause 11.1). The IChemE justifies its abandonment of the 'Engineer' (cf. Red Book 2nd Edition) as being insufficiently widespread for use in the process industries, while recognising that most clients appoint one of their own staff as a project manager. This development is a repeat of the revision made in the Green Book 2nd Edition in 1992. However in the Green Book, the Guide Notes state that the Institution recognises it may be difficult for an employee of the client to act impartially;

whereas the Red Book 3rd Edition expects this appointee to be 'impartial' as between the Purchaser and the Contractor (clause 11.1(d)).

There is an apparent contradiction here between Red and Green Books which highlights the difficulty of clients' appointing their own staff to administer the contract. Furthermore, as the Project Manager is given specific responsibilities to determine judgements between the Contractor and Purchaser (some of which are final and binding and not subject to the review of arbitration or court proceedings – see, for example, clauses 12.5 (removal of Contractor's supervisory staff) or 42.1 (suspension of the works)), the need to ensure non-partiality is important (not least of all for the Contractor!).

The Project Manager is not expected to supervise the construction of the works (as in the ICE's Engineer) and although he may appoint an assistant (the 'Project Manager's Representative') this person should not be considered as an equivalent of the Resident Engineer or Clerk of Works. Clause 11.4 gives the Project Manager's Representative authority to condemn any design, workmanship or materials but, unlike the ICE Conditions, there are no terms giving the Project Manager the right of inspection before covering. In this respect the Project Manager is more of a contract administrator than technical supervisor.

Variations to the works can be made in the course of the works by the Project Manager and are generally defined as any alteration in the Plant, method of working, programme of work or to the type or extent of the works, which is an amendment, omission or addition thereto (clause 16.1). However unlike other standard forms, the Contractor is given the right of objection to the ordered variation provided that appeal is lodged to the Project Manager within 14 days of receipt of the instruction (clause 16.6). In this case the Project Manager is expected to re-consider his instruction and either withdraw it or insist upon it. Any continuing dispute over the enforcement of the variation is referred to the Expert for a final decision (clause 45).

Thus, the instruction of variations is not seen as a totally unilateral process. Clause 16.4 allows the Project Manager to 'instruct' the Contractor for assistance in the preparation of variations (which typically might be quotations for the work, or technical advice on specific matters concerning the work). Similarly if the Contractor is of the opinion that a variation to the works is required, a written proposal may be submitted to the Project Manager for his consideration (clause 17.1).

This reflects some of the IChemE's philosophy of co-operation in drafting the contract. The introductory notes to the Red Book state:

"...the parties should co-operate to achieve the mutual objective of a successful project rather than regarding the contract as the basis for an adversarial relationship..."

However, there is still much within the IChemE Red Book which is in the 'arms-length' mould of contract drafting and this expectation is somewhat hopeful in the absence of a suitable determining relationship. Furthermore, the IChemE's Purple Book[13] specifically refers to the Red Book as an arms-length contract and acknowledges that in these type of contracts: *"...there will always be a potential confrontation element below the surface in the project management relationship"*.

Again here, there is a contradiction between the Red Book and other IChemE publications. More specific guidance is required on the establishment and maintenance of an appropriate relationship to govern the contractual conditions. Rather than attempt to determine the relations through the contractual terms or insert a hopeful note in the introduction, specific alignment of the conditions of contract with a determining relationship is required. Given this, the specific risks associated with contracting will either be obviated (in the case of competitively established collaboration), or fully covered by the express terms of the contract (where arms-length adversarial relations prevail). The IChemE's 'mix and match' approach to contractual relations is unlikely to work in the commercial world of construction.

Some specific points worth noting about the terms of the Red Book include:

- The 3rd Edition has been modified by the IChemE to recognise laws, regulations and codes of practice other than those of the

UK. However it would appear that the contract still requires considerable specific modification for international use, as is suggested in Guide Note LL.

- Despite being published after the introduction of the Construction (Design and Management) Regulations 1994, the 3rd Edition makes no specific account of them in its conditions and only refers to them in passing (Guide Note L).
- Schedule 9 and Guide Note R suggest that the Purchaser may decide on an arbitrary percentage figure of the Contract Price to be inserted as the value for liquidated damages. Where such a figure genuinely represents the Purchaser's pre-estimate of the likely loss in damages resulting from delay, then this practice can be considered sufficient. However, in certain circumstances, such practice could be considered to be the levy of 'penalties' rather than liquidated damages, which would result in the damages becoming unrecoverable. Contrary to IChemE guidance, the client is advised to take specific legal advice before inserting anything other than a genuinely pre-estimated figure of specific value.
- The IChemE refers to *force majeure* events in clause 14. These are defined as circumstances beyond the reasonable control of either party which prevent or impede the due performance of the contract. Although in English Law there is no statutory concept of *force majeure*, the IChemE (unlike the JCT) has circumnavigated this by allocating a specific definition to the term and listing those events which constitute it (clause 14.1). This is acceptable practice and, as such, would be supported by the courts. The half-hearted attempt by the JCT to include for *force majeure* is, in reality, meaningless.
- As with the GC/Works/1 discussed in Chapter 11, the Red Book makes provision for regular progress reviews between the parties (clause 29).
- The 3rd Edition of the Red Book has introduced Alternative Dispute Resolution (ADR) to the IChemE dispute resolution process. There is now a choice for the aggrieved parties between reference to an Expert (clause 45) or Mediation followed by Arbitration (clause 46).

Summary of the IChemE Red Book

In summary, the IChemE Red Book is a lump sum D&B performance contract with an emphasis on the performance of the completed works, rather than the means by which they are to be constructed. It contains particularly well-specified provisions for the testing and commissioning of the works, which is appropriate since this is what the contractor is measured against and reimbursed for. This review has only served as a brief introduction to the contract, nevertheless it has raised some concerns which the IChemE have not addressed well. These include:

- the contradictory comments regarding the appropriate relationship to accompany this contract;
- the IChemE's failure to recognise the effect of the CDM regulations;
- the question of the legality of the liquidated damages.

In a similar way to other D&B contracts, the balance of power rests firmly with the specification and performance criteria set by the client; should these be inadequate the client will pay dearly for any costs of change. As the contract is performance-based and non-prescriptive, it follows that the risks and balance of power are better defined and more appropriately apportioned. In effect the IChemE has made contracting simpler by giving clarity to the discharge of contractual requirements and duties.

D&B Contracts: General Conclusions

This chapter has examined the design-build process and its contractual mechanisms to demonstrate a number of key learning points concerning the balances of power and risk. Despite the obvious differences between the ICE, JCT and IChemE design–build contracts, similar mechanisms are at work. The principle of *single point responsibility* provides a greater degree of resolution and clarity to the questions of risk allocation and contractual responsibilities than that offered by the sequential contracts.

The critical fulcrum-point is clearly at the point of tender and award. It is here that the client relinquishes power to the contractor to undertake the works. It therefore follows that the client is dependent on well-defined specification and supporting contract documents. It also follows that it is in the client's interest to exert as much *pre-contractual leverage* as is practicably possible in order to procure the works most favourably. After the contract has been awarded the client has little or no leverage in the contract and thus will need to consider other means of commercial leverage or negotiation to make late changes or re-direct the works.

The question then follows: which of the D&B contracts is most appropriate? The answer is clearly dependent on the contingent circumstances of the commercial transaction. However it is clear that, assuming the client has specified the works competently and that it has been able to effect the most economically advantageous deal through the competent application of its pre-contractual leverage, any of the three contracts will be suitable. Under the mechanism of single point responsibility the contractor is left to its own devices to deliver the works to requirement.

If there is some doubt about the competence of the client (or its profesional advisor) to adequately specify the works and/or the competence of the contractor to meet the specified demands, then the ICE contract would appear to be the most robust form of contract to use. In the absence of these uncertainties, the IChemE Red Book, with its primary focus on the performance outcome, would seem most appropriate.

Further Reading

The discussions in this chapter have not been produced as a legal critique of the provisions of D&B contracts; they only serve as an introduction. Unfortunately, space has precluded greater analysis of many of the finer points of these contracts. The following section lists some of the more detailed texts available to readers (at the time of this publication being written) who may wish to pursue more detailed studies of D&B contracting.

For further reading on the ICE Design and Construct contract, the reader's attention is drawn to: Brian Eggleston's *'The ICE Design and Construct Contract: A Commentary'* published by Blackwell Scientific Publications, Oxford (1994).

For further reading on the JCT'81, the reader's attention is drawn to: the JCT Practice Notes CD/1A and CD/1B; D. Chappell & V. Powell-Smith's *JCT Design and Build Contract*, published by Blackwell Scientific Publications, Oxford (1993); D. Janssens' *Design-Build Explained*, published by Macmillan, London (1991); and D. Turner's *Design and Build Contract Practice, 2nd Edition*, published by Longman Scientific & Technical, Harlow (1995).

At the time of this publication, there has been very little available literature to comment on the provisions of the 3rd Edition of the IChemE Red Book, as it has only been in use one year. Nevertheless the contract itself has ample introductory and guidance notes for the newcomer.

Chapter Notes

1. Cox A. & M. Townsend (1998) *Strategic Procurement in Construction*, Thomas Telford, London.
2. Refer to: The Business Roundtable (1983) *More Construction for the Money: Summary Report of the Construction Industry Cost Effectiveness Project*, The Business Roundtable, New York.
3. Bennett J., Pothecary E. & G. Robinson (1996) *Designing and Building a World-Class Industry*, Centre for Strategic Studies in Construction, The University of Reading.
4. The authors' analysis of the financial performance of four of the top UK contractors in 1994 and 1995 shows that Return on Sales (i.e. Net Profit/Turnover ratio) varied between 0.8% and 2.9%.
5. Refer to: Eggleston B. (1994) *The ICE Design and Construct Contract: a Commentary*, Blackwell Scientific Publications, Oxford.
6. Broadly translated, the *contra proferentum* rule means: the words of a document are to be construed against a person seeking to rely on them.
7. According to: Murdoch J. & W. Hughes (1996) *Construction Contracts: Law and Management, 2nd Edition*, E&FN Spon, London.
8. Refer to: JCT Practice Note 20 *Deciding on the appropriate form of JCT Main Contract* (August 1993).
9. Janssens D. (1991) *Design-Build Explained*, Macmillan, London.
10. Refer to: Turner D., (1995) *Design and Build Contract Practice, 2nd Edition*, Harlow, Longman Scientific & Technical, p. 38.
11. Bennett J., Pothecary E. & G. Robinson (1996) *Designing and Building a World-Class Industry*, Centre for Strategic Studies in Construction, The University of Reading.
12. This case study was first reported in an article titled 'Driving down the costs of building a drug plant' in *Manufacturing Chemist*, March 1996.

13. Wright D. (1993) *An Engineer's Guide to the Model Forms of Conditions of Contract for Process Plant*, IChemE, Rugby.

Chapter 7:

Minor Works Contracts

Introduction:

In the brief introduction to the Minor Works forms of contract in Chapter 4, it was noted that these contracts do not constitute a different contracting method to any of the common methods mentioned so far (Chapters 5 and 6). Indeed the only differentiating factor appears to be the size (that is, the magnitude) of the works contract in pure monetary terms; other than this the construction process is similar to the sequential contracting method.

This accepted, there follows a number of pertinent questions to consider:

- are minor works contracts really necessary, or are the other contract forms adequate?
- why should the contract sum become the differentiating factor?
- if the monetary value is not the most appropriate deciding factor, what criteria should we consider to decide whether a minor works form is more preferable than a 'fuller' standard form of construction contract?

In essence this chapter asks whether minor works add any particular value and, if so, how should 'minor' be defined? To answer these questions, we first need to consider why minor works forms were introduced to the industry and what is the relevant structure of power behind them.

Why Were 'Minor Works' Necessary?

According to commentators on the ICE Conditions of Contract for Minor Works[1], the need for a minor works form was: *"...prompted by a desire to reduce the amount of paperwork that was being created on most major contract."* and that, furthermore: *"Experience had shown that quite a large number of people were incapable of understanding and completing correctly the Appendix to the Form of Tender for the ICE 5th Edition."*

Aside of the immediate doubts this raises for the ICE's main contract, these comments would appear to support a need to develop a shorter and simpler standard form of contract. Such a contract would require an appropriately smaller proportion of administrative on-costs, in keeping with the potentially smaller prime costs in the contract sum. This, then, would appear to be the intentions behind the ICE Minor Works form.

How, Then, Should We Define Minor Works?

Having accepted that, in principle, there is a need for simpler and shorter forms of contract under certain (albeit contingent) circumstances, it follows that it is necessary to know what those contingent circumstances are. Once known, the project sponsor knows when minor works contracts are appropriate to be used. To help decide this, it is first necessary to establish what constitutes 'minor works'. Having a handle on their definition, will lead the prospective user to know if a minor works contract is appropriate, or not.

The Guidance Notes to the ICE Minor Works contract (2nd Edition) suggest that the contract is intended for works:

- where the potential risks involved are small for both parties;
- that are of a simple and straightforward nature;

- where the contract value does not exceed £250,000; or
- where the contract duration is less than six months.

The guidance offered by the JCT for its Intermediate Form of Contract (the IFC'84) is equally vague; it suggests the IFC'84 should be used where the proposed works are:

- of a simple content involving normally recognised, but basic, industry skills or trades;
- without complex service installations or other specialist works from other trades disciplines; and
- adequately specified and/or billed prior to invitations to tender.

In other JCT guidance notes[2], it is suggested that the IFC'84 is appropriate for works of not more than £280,000 (at 1992 prices). Since the same notes suggest the JCT Minor Works contract (the MW'80) is suitable for works of less than £70,000 in value and that the JCT Jobbing Contract is suitable for works less than £10,000 in value, one is left unclear as to what is the definition of minor works.

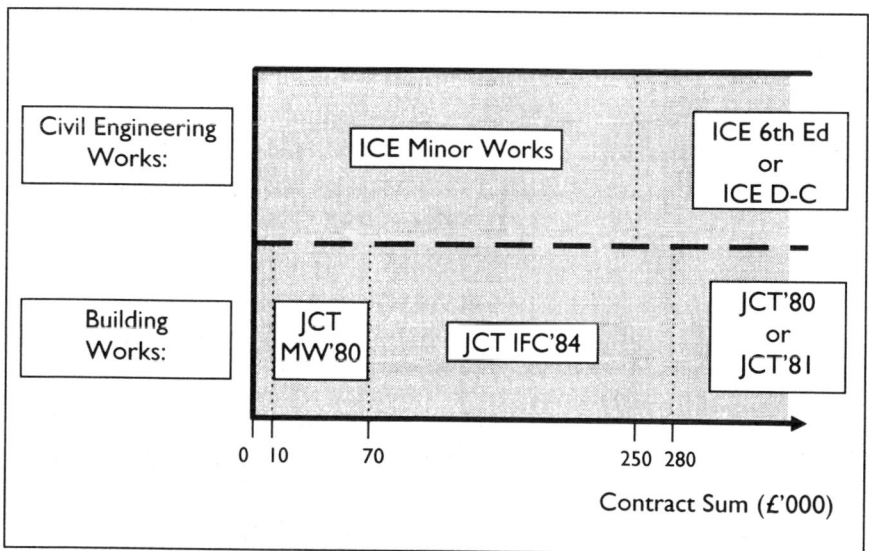

Figure 7.1: The Financial Bounds of Minor Works Contracts.

Thus, if we consider the value of the contract sum alone, there is a clear basis of guidance on the appropriateness of minor works contracts (as illustrated in Figure 7.1). However is this sufficient? Should the contract sum be the only indicator of minor works? What about the issue of *risk* to the contracting parties? Surely, irrespective of the simplicity or value of the contract, some construction works can develop into unplanned costly events.

There is no direct relationship between risk and the value of the contract. The same generic risks will still be present (albeit of a smaller magnitude) and the project-specific risks will just be different: there is no telling whether they are larger or smaller and/or less or more likely to occur. Moreover the construction works will be required for different purposes; some may be required for assets which are 'residual' to the client's business, while others are required for more important 'critical' assets. Thus the defining criteria for 'minor works' need to consider the strategic importance of both the works and the asset which is undergoing construction[3].

Clearly, where the risks within the construction works are high and the likelihood of an adverse impact on the business is also high, a 'minor works' contract may not be the most appropriate form of contract to use, without some other controlling mechanism in place (such as a governing relationship). Indeed, one commentator has stated: *"As with all briefly drafted and simply worded forms of contract on a subject as complex as works of construction there is some lack of legal precision. The price of brevity is paid with uncertainty and the minor works form does leave a few questions answered"*[4]. Furthermore, regarding the JCT Minor Works contract (the MW'80), one commentator has written: *"Its simplicity is deceptive....there are pitfalls for the unwary"*[5].

Does this mean that all minor works forms of contract lack precision and should not be advocated? Probably not; but there do appear to be some question-marks over the utility of a standard form of contract, written for generic use, but which only covers certain risks lightly. Even smaller works carry significant risks on a project-specific basis; a point that has been already laboured. Does this therefore mean that a minor works contract can only ever be 'fit-for-purpose' when written on a 'one-off' bespoke basis? If this is the case, it is hardly going to address the need for less expensive administrative on-costs for smaller works!

Thus, there are few clear guidelines or conceptual frameworks that can be offered to the practitioner here. Clearly, the application of simpler and shorter forms of contract can only be considered in the context of the contingent circumstances of the transaction and the specific type of works to be contracted. It will also be clear to the reader that a compromise may be required between project-specific contractual instruments and cost-effective administration.

For example, a client requiring a small number of building works ranging up to £350,000 per contract might be forgiven for despairing at the range of JCT contracts it needs to be familiar with. The cost inefficiencies of being familiar with four different forms of contract, for fairly small workloads, do not seem commercially pragmatic or viable. In Case Study 5 at the end of this chapter, we consider how Whitbread plc has used the IFC'84 on a diverse range of construction projects as a 'company-standard' approach.

The issue of the magnitude of works and the relative costs of transaction has been presented to the industry as a compromise in the form of the available standard forms of contracts for minor works. The remaining sections of this chapter examine the provisions of two of the most common minor works forms of contract (the ICE contract and the IFC'84) and the power structure inherent within them. In 1997, the Institution of Civil Engineers published the NEC Small Works contract and this is briefly mentioned in Chapter 8. Other forms of contract do exist, such as the JCT MW'80 or the JCT Jobbing Contract, but space has precluded any detailed mention of them here.

ICE Minor Works, 2nd Edition

The ICE Conditions of Contract for Minor Works is not like the ICE 6th Edition in terms of drafting or parallel clauses. Thus familiarity with the ICE 6th will not enable users to fully interpret the ICE Minor Works form, even though many of the principles are the same and, in effect, the parties in the contract perform similar roles. Essentially the balance of risk between Employer and Contractor remains the same, with the Engineer administering the contract yet not being party to it.

The ICE Minor Works form provides for a variety of methods of reimbursement:

- lump sum;
- admeasurement using a Bill of Quantities;
- valuation based on a schedule of rates;
- valuation based on dayworks; and
- cost-plus reimbursement.

Provision is made in the Appendix to the Conditions of Contract for the choice of reimbursement to be specified and, usually, this is either lump sum or admeasurement. In the case of lump sum, interim payments are valued on a judgement of the percentage of work achieved. Whereas this may be a simple administrative task (albeit subject to the Engineer's opinion) this can become problematic when assessing claims for additional payment. Alternatively the admeasurement basis of 'measure and value' can become administratively burdensome on works of low value. If a Bill of Quantities approach is required the rules of measurement need to be defined at the pre-tender stage; as these have not been included in the ICE form. A general assumption that the ICE contract will follow the typical CESMM3 will not be recognised in law.

There are some specific issues regarding the use of the ICE Minor Works form which are worth particular mention. These include:

- the ability to elect design responsibility to the Contractor for part of the works. This provides flexibility to apportion the design to the most competent player, which is particularly appropriate for minor works where the intention is to keep the work requirements simple. As with the ICE 6th Edition, design liability is limited to reasonable skill, care and diligence. Partial Design contracts were discussed in greater detail in Chapter 5.
- The ICE Minor Works 2nd Edition has included a clause to cater for the Construction (Design and Management)

Regulations 1994. Again, like the ICE 6th, this is a potential area of concern for the administration of the contract. Clause 13.3 allows the Planning Supervisor to make alterations to the Health and Safety Plan, which has the effect of an Engineer's Instruction to vary the works and give cause for additional payments and/or extensions of time.

- There are potential problems with the application of extensions of time and/or liquidated damages when 'part of the works' are completed. The definition of 'part' is not precise; it appears to refer both to parts defined in the schedule and to parts taken into use/possession independently of the remainder of the works.

- The ICE Minor Works form retains the ICE 'retention of money' provisions for ensuring satisfactory performance and completion of the works. The intention is that, in retaining a small percentage of the payment, the Employer may grant 'substantial completion' for the works on the understanding that the outstanding items will be wholly completed in the period of time allotted (the 'defects correction period', formerly known as the 'maintenance period'). The retained payment is held as an incentive for the Contractor to fulfil his obligations, upon which it is released and paid to the Contractor. However the Guidance Notes for the ICE Minor Works recommend a rate of retention between 2.5% and 5% which, even at the recommended maximum value of the works (£250 000), is only £6250 to £12,500. In some circumstances this will not represent sufficient incentive for the Contractor to re-mobilise and return to site in order to finish off the outstanding work items (which are often the most time-consuming).

Summary of ICE Minor Works contract

The above points serve only to highlight some particular issues relating to the administration of the ICE Minor Works form. For the reader who wishes to study the provisions in greater detail, reference is recommended to Cottam and Hawker[1], although it should be noted that this is written for the first edition and is therefore now out of date.

Like its parent contract, the ICE 6th Edition, the minor works contract is over-reliant on the role of the Engineer. Once again this contract removes power and control from the client and gives it to a third party 'professional' who is not privy to the contract. Thus, there are clear dissonances between: (1) the Engineer's powers and the liability for his actions and (2) the Engineer's impartiality and the service to his paymaster.

While the ICE Minor Works contract is flexible and versatile for the needs of smaller projects, it does carry a number of anomalies and potential issues of concern and these have been outlined briefly. Overall, although the contract apportions a fair balance of risks, the client is still placing considerable power in the hands of the Engineer with no apparent recourse. This imbalance should be redressed in the Engineer's terms of engagement.

The JCT IFC'84

The full title of the IFC'84 is *The Intermediate Form of Building Contract for Works of Simple Content, 1984 Edition,* which provides an immediate indication of its content and intentions. It has very similar provisions to the JCT'80 and was introduced to fill a gap between this contract and the MW'80. Thus the contract is designed on the JCT'80 model albeit drafted in a shorter format that is easier to read. Despite the similarity, the clauses do not run parallel with the JCT'80; thus familiarity with it will not enable the user to administer the IFC'84 with competence.

The recommended range of construction projects has already been detailed (£70,000 to £280,000). However in a survey conducted by the JCT in 1987/88, approximately 49% of clients used the IFC'84 for works between £250,000 and £1 million, while a further 20% used it for works of between £1 million and £5 million[6]. It has already been noted that the JCT's attempt to produce a contract for every type of arrangement hardly seems to be practicable or to have the client's best interests at heart. Thus it is probably this factor which has led to the wider-than-recommended use of the IFC'84.

In drawing a comparison with the JCT'80, the following points should be noted:

- The IFC'84, like the Standard Form, represents the 'sequential' approach to procuring building works in that it separates out the design and construction elements of the works and has a professional third party preside over the contract administration (the architect/contract administrator, as defined in Article 3). There is no comparable 'intermediate' ICE contract; if anything, the IFC'84 is most similar to the ICE Minor Works form.
- The IFC'84 does not provide for any design by the contractor either in part or in full. Similarly there is no Contractor's Designed Portion Supplement to accompany it.
- Unlike the JCT'80, there is just one standard form of the IFC'84 available. It is suitable for both private and local authority use, with reimbursement either on a "measure and value" basis (With Quantities) or on a lump sum basis (Without Quantities). Distinction between the two is to be made in the invitation to tender.
- The insurance provisions for the IFC'84 (clause 6) are exactly the same as those for the JCT'80.
- There is no requirement on the contractor to submit a 'master programme' (as in the JCT'80), indeed there is no mention given to the timing or progress of the works between commencement and completion, other than: *"The Contractor shall carry out and complete the Works in a proper and workmanlike manner..."* (clause 1.1) and *"The Contractor shall...regularly and diligently proceed with the Works and then shall complete the same on or before the Date of Completion..."* (clause 2.1). It has been argued that, although there is no contractual requirement for a programme, if one is produced by the contractor, his failure to keep to it could be viewed as a failure to 'regularly and diligently proceed with the works'[7]. However the absence of programme raises other issues too: (1) it is very hard for the employer to monitor/control the progress of the works; and (2) it makes claims for extensions of time both harder to justify and harder to evaluate. Thus its

absence is clearly to the client's detriment, particularly where time is a major determinant (driver) of success.

- Perhaps the most noticeable difference between the IFC'84 and the JCT'80 is the absence of nomination in the IFC'84. Instead, as well as the usual 'domestic' sub-contractors, the IFC'84 provides for subcontractors to be 'named' (clause 3.3). This can be achieved either by naming in the documents or through the presence of a provisional sum, to be used (instructed) in the course of the works. The procedures are not unlike nomination and can create a substantial quantity of documentation. Furthermore there seems little point in naming the subcontractor or supplier if, as the IFC'84 recommends, the work is not of a specialist nature (especially as nomination often leads to single-sourcing and dependency relationships). If the works are so special or complex as to require a named supplier, then the IFC'84 is not the form of contract best used.
- Finally, in the event of a dispute, the IFC'84 refers the parties to arbitration (the JCT Arbitration Rules) which seems to be an inappropriately heavy-handed method of dispute resolution for 'works of a simple content'. In this instance adjudication would appear to be more preferable, even preceded by mediation or conciliation as a short and relatively inexpensive alternative. Further discussion on conflict management and the merits of various dispute resolution procedures is made in Part D of this book.

Summary of the IFC'84

To summarise, the IFC'84 was introduced for building works of a simple content, generally between £70,000 and £280,000 in value. Practice has shown however that the contract has far wider application than this, as it represents a simpler and more welcoming version of the JCT'80. There are specific concerns related to its use which include:

- the lack of control on timing and progress (thus making it an inappropriate form of contract when time is an important driver for the works or its environment);
- the inappropriateness of 'naming' specialist suppliers;

- the inappropriate provisions for 'simple' dispute resolution.

The structure of power in the IFC'84 is designed to favour the contractor. The client, as with so many of the standard forms of contract, is virtually powerless and depends on the capabilities of the architect/contract administrator. However, as there are so few third party controls written into the contract, the balance of power favours the contractor. The contractor may do what he wishes to complete the works and, should his progress be changed or re-directed, the client will be expected to pay for the interruption.

For further information on the IFC'84, recommended texts include:

1. JCT Practice Note IN/1 *Introductory Notes on the Intermediate Form of Building Contract, IFC'84* (Revised September 1994);
2. Chappell D. (1995) *Understanding JCT Standard Building Contracts, 4th Edition*, London, E&FN Spon;
3. Jones, N.F. & D. Bergman (1990) *A Commentary on the JCT Intermediate Form of Building Contract, 2nd Edition*, Oxford, BSP Professional Books.

Case Study 5: Whitbread plc

Whitbread plc is one of the UK's leading leisure companies. As well as a operating renowned brewery, it owns and manages some of the country's most popular pubs, restaurants, hotels and leisure facilities. The group has a large property portfolio of well-established branded facilities which are currently valued at approximately £2.3 billion, including:

- Over 290 Beefeater Restaurants and Pubs;
- 20 David Lloyd Leisure sports clubs;
- Approximately 1700 pubs and pub restaurants (including names such as Brewers Fayre, Wayside Inn and Hogshead);
- Over 1500 drinks retailing outlets including: Thresher, Bottoms Up, Wine Rack and Drinks Cabin;
- Numerous restaurants, including Pizza Hut (450 outlets), TGI Friday's (23), Cafe Rouge (25), Bella Pasta (90) and several other ranges of concept restaurants;
- Over 20 major brewing sites for well-known beers such as Boddingtons, Stella Artois, Heineken and Murphy's;
- 146 Travel Inn hotels and a further 32 Marriott hotels.

Continued

As a result of this diverse and sizeable asset-base, Whitbread's spend on property development is high at approximately £400 million p.a, of which approximately 60% represents new-build construction activities.

Whitbread has pursued a number of 'fit-for-purpose' procurement strategies which have been driven by the needs of the business for speed to market and cost-reduction. Methods to service these drivers have included off-site modular construction of 'pods' (such as toilet blocks) and other value engineering techniques.

Since mid-1996, Whitbread has engaged 'partnering' agreements with a small group of their preferred suppliers in order to achieve these objectives. The partnering agreements are built around the use of the JCT IFC'84 form of contract, as it is seen to be a simple and yet versatile building contract. Larger contracts, such as the £25 million new build Marriott hotels, are procured through the JCT'80, but for the rest of Whitbread's workload (the majority) the IFC'84 is deemed to be sufficient. In fact Whitbread would prefer it if the contract was not referred to and the works were carried out on the basis of the relationship it is developing with its suppliers.

This relationship is based on a *coincidence of interests* between Whitbread and its suppliers – i.e. both parties get what they want from the relationship. For Whitbread this means quick on-site mobilisation combined with value engineering exercises and fixed-price profit margins. The value engineering is based on the optimal balance of functionality and lowest cost. All savings identified and realised are retained by the client. The suppliers know that their contributions to profit and on-cost remain untouched and that, should they continue to satisfy Whitbread's needs, there is a continuity of workload.

The benefits of this approach are already emerging. One example from new build TGI Friday restaurants has suggested that 40% reductions in the capital cost and 50% reductions in the time to completion are readily achievable through these procurement methods.

This is not to say that these benefits can be accrued directly from use of the IFC'84 contract alone. The IFC'84 has been a contributory factor to the 'partnering' relationship that has enabled Whitbread to achieve a higher level of service from the construction supply-base.

Source: Whitbread[8]

Chapter Notes

1. Cottam G. & G. Hawker (1991) *ICE Conditions of Contract for Minor Works: A User's Guide and Commentary*, Thomas Telford, London.
2. JCT Practice Note 20 *Deciding on the appropriate form of JCT main contract (revised August 1993)*, RIBA Publications Ltd, London.

3. Refer to: Construction Round Table (1997) *Strategic Risk Audit Methodology and Guide for Construction Procurement*, produced by Andrew Cox Associates, 147 Luddington Road, Stratford-upon-Avon, UK.

4. Egglestone B. (1992) *Liquidated Damages and Extensions of Time*, Blackwell Scientific Publications, Oxford.

5. Chappell D. (1995) *Understanding JCT Standard Building Contracts, 4th Edition*, E & FN Spon, London.

6. Joint Contracts Tribunal (1990) *The Use of Standard Forms of Building Contract*, RIBA Publications Ltd, London.

7. Jones N. F. & D. Bergman (1990) *A Commentary on the JCT Intermediate Form of Building Contract, 2nd Edition*, BSP Professional Books, Oxford.

8. The information presented in this case study was taken from *The Whitbread Briefing Book, Issue Nine* (1997) and discussions with Whitbread staff in the course of the author's research. For further discussions on the merits of partnering and other relational approaches see: Cox A. & M. Townsend (1998) *Strategic Procurement in Construction*, Thomas Telford, London.

Chapter 8

The NEC

Introduction

The New Engineering Contract (NEC) has been given a specific chapter of its own in this book as it is quite different from any other standard form of contract available in the industry. Strictly speaking the title NEC refers to a 'family' of contract documents and not just one specific contract. It is still a relatively new phenomenon and subject to considerable on-going debate amongst academics and practitioners alike. This chapter seeks to draw on some of these debates to examine the inherent structure of power and the strengths and weaknesses of the contract family. Although the chapter concentrates on the Engineering and Construction Contract (ECC) as the main NEC contract, much of the examination can be extended to other forms of contract in the NEC collection.

The ECC forms part of the second edition of the NEC published in 1995 by Thomas Telford Services Ltd for the Institution of Civil Engineers. It has been marketed as the only fully 'Lathamised' contract; however to call it a mere form of contract is to misunderstand its mechanisms completely.

The ECC has kept hold of the NEC family name for good reason. It is referred to as a family because it comprises a suite of compatible (i.e. interlocking) contracts suitable for a variety of uses and contracting methods, as well as a number of secondary options for contractual flexibility. This compatibility extends beyond the main [client-contractor] contract to other contractual links in the supply chain.

The launch of the NEC created a stir because of its revolutionary format. Unlike any of its contemporaries, the NEC was drafted from a clean sheet based on 'good practice' principles of project management. Newcomers to it might be a little uneasy about some of its terminology and contractual language; but once the veil is lifted it really is very simple; the basic contracts are only a few pages long (despite the 2.8 kg pack they come in).

According to the drafters ('The NEC Working Group')[1], the principal objectives were to create:

- flexibility,
- clarity and
- a stimulus to good project management.

Whether the ECC has succeeded in achieving these goals is still open to debate. Certainly corroborators of the NEC[2] claim the implementation of the contract has been a success and that there are case studies to support their claim (BAA, National Power, Anglian Water, Yorkshire Water and others). However some scepticism remains, which is hardly surprising for a deep-rooted traditional industry. One commentator has gone as far to suggest that the only reason for its success is that the parties 'were out to make it succeed', which suggests that this is not the usual intentions of business in construction. Possibly the more likely reason for the industry's reserve to using the NEC (*c.* 500 users to date in the UK at the time of publishing) comes from a lack of familiarity with its new methods and concept.

One recent article examined 28 of the earliest NEC-procured projects[3]. Although it had mixed conclusions some scepticism must be given to the validity of such early 'evidence'. These projects have been managed by supportive clients who appoint their most able project-managers with their favoured contractors to monitor the NEC carefully as it undergoes its trial period in the organisation. Empirical evidence drawn from such peculiarly false circumstances must be treated with extreme caution. The only true test will be over thousands of construction projects of varying size and complexity among a vast range of commercial circumstances.

Therefore 'success' for the NEC needs careful consideration. Performance, as the determinant of success, needs to be quantifiable in order that the practitioner knows which criteria apply and how they are measured. Thus the NEC may have successfully achieved its three principal objectives, but whether this has been completely *fit for purpose* remains untested as very little performance data has been published.

If the new procedures create an administrative burden, as some concerns have raised[4], careful analysis is required to determine whether this detracts from the original intent (purpose) of the works. Reliance on procedures to shape relationships have, throughout the history of business, bureaucratised practice and deflected the focus away from performance. Whether this is the same in the case of the NEC remains to be established, but clearly there is a focus on *procedure* rather than performance.

Latham's support of the NEC

In his review of the construction industry[5], Sir Michael Latham suggested certain principles which should be found in any modern form of contract. To the surprise of some, Latham endorsed the NEC (1st edition, published 1993) as possessing virtually all those preferred attributes. To improve the first edition, he recommended seven amendments (see Figure 8.1) but these have yet to be broadly incorporated in the ECC. Whether this new edition, or any other contract which resembles Latham's 'modern contract', will provide the transformation of culture that the industry is looking for remains doubtful[6]. As has been suggested elsewhere[7], a more fundamental structural re-organisation of the industry may be necessary based on fit-for-purpose supply *relationships*. The key question is whether there is a home for the NEC family in here or not.

Latham's recommended alterations to the NEC	Date when implemented
• Change name to *New Construction Contract.* • Make the Trust Fund clauses mandatory. • Review payment periods. • Make the duty to undertake the contract in a spirit of mutual trust and co-operation a mandatory clause. • Expressly state that no core clauses should be amended by either party to the contract. • Publish standard terms for consultants and adjudicators and publish standard tender documentation. • Publish a minor works contract.	• Not implemented. • Not implemented. • Shortened in ECC. • Implemented in 2nd Edition. • Not implemented. • Not implemented. • Published in 1997.

Figure 8.1: Latham's Recommendations on the NEC.

Key Characteristics
(1)   Family Of Contracts
The NEC family is a suite of contracts not just one rigid form. Flexibility is the key, as discussed in sub-section (3) below, with six 'Main Options' providing a variety of contracting methods:

Option A. Priced contract with activity schedule;
Option B. Priced contract with bill of quantities;
Option C. Target contract with activity schedule;
Option D. Target contract with bill of quantities;
Option E. Cost reimbursable contract;
Option F. Management contract.

Other contracts in the family include a sub-contract, a contract for professional services, a plant installation contract and a contract for the provision of an Adjudicator. These have been drafted on the same principles as the ECC and as such interlock using the same terminology and type of language. Thus the sub-contract really can be 'back to back' with the main contract. (Under clause 26 of the ECC, the Contractor is required to have all forms of subcontract approved by the Project Manager, unless the NEC form of sub-contract is adopted, thus the selection of conditions is openly discussed and the assignment of certain risks can be controlled.) It should also be noted that the NEC Subcontract does not employ 'pay-when-paid' clauses, unlike the FCEC Blue Form of Subcontract.

Case Study 6: Use of NEC at Eskom

Eskom is the world's fifth largest electricity utility with 37,000 MW capacity and more than 250,000 km of transmission grid. It principally operates in South Africa, but has hydro-electric operations in other African states.

In 1988 Eskom began to review its contracting and project management practices. It had been using a set of antiquated bespoke forms of contract drafted back in 1939 which were updated on an occasional basis. However these contracts were not fulfilling all of Eskom's business objectives. After some investigation, it was decided that Eskom's contracts needed to:

- provide good project management guidance;
- account for a variety of contractual and commercial options;
- employ everyday language capable of being understood by a diversely cosmopolitan society;
- be suitable for international use.

After examining other European standard forms of contract (including FIDIC) Eskom found none that suited its specific needs. As a result it proceeded to draft a bespoke contract based on the South African ICE Conditions of Contract, 6th Edition. However this was shortly abandoned when the ICE's Consultative Edition of the NEC was introduced to the company in 1991.

Since then, Eskom has been the world's single largest user of the NEC. Following the publication of the Consultative Edition, Eskom recognised the need to have a Minor Works form of contract and proceeded to draft its own. This has now been used on more than 2000 occasions as part of Eskom's rural electrification programme which runs at approximately R2000 million p.a. The NEC Working Panel has been glad of Eskom's support and many of the changes in the 2nd Edition have been as a result of feedback from Eskom's experiences.

Source: Baird[8]

One of Latham's recommendations was to publish a simpler and shorter minor works form which was finally published in 1997. This was based on the bespoke form in operation in South Africa's electricity generating company, Eskom (see Case Study 5), although it has already received some criticism as being *"...nasty, brutish and short..."* in Hobbesian terms.

Why the need for a short form?

One of the NEC's authors, Professor John Perry, has observed that small contractors lack the expertise to comply with the full NEC procedures[9]. Others have suggested that the NEC is less suitable for dealing with smaller firms that have no experience of planning or estimating. It also follows that proportionally the contract administration is a greater burden on works of smaller value; smaller firms may lack the resources to operationalise the NEC provisions. This therefore provides sufficient reason for the adoption of a minor works form into the NEC family.

However, as Perry has asked: how can 'minor works' be defined? A simpler form of contract may choose to omit certain clauses because they cover risks of smaller magnitude, but which those omissions should be would be difficult to determine except on a specific project-by-project basis. Thus the inference is that a fit-for-purpose minor works form needs to be bespoke each time (except on repeat works), which is hardly going to reduce the costs of paperwork and administration which minor works forms of contract are designed to do. The counter view suggests that standard forms of contract for small works (such as the NEC's) will be sub-optimal and fail to be *fit-for purpose*.

(2) Terminology

One of the NEC's most distinguishing features is its language. Figure 8.3 (overleaf) illustrates some examples of the changes from traditional terminology. Of most note is the former role of the Engineer which has now been split between Project Manager, Supervisor and Adjudicator. The Project Manager is the client's representative whose *"...role within the ECC is to manage the contract for the Employer with the intention of achieving the Employer's objectives for the completed projects"* (ECC Guidance Notes). There is no obligation to act fairly or impartially: the Project Manager is clearly acting for the client. The ECC places considerable authority in the hands of the Project Manager; it assumes that this person has the Employer's full authority to make decisions and carry out the required actions which, in practice, may be difficult in the delegation of authority in certain organisations (particularly the public sector or other organisations with rigid standard operating practices).

NEC	Other ICE Contracts	JCT Contracts
Project Manager	Engineer, or Engineer's Representative	Architect, or Quantity Surveyor
Supervisor	Engineer's Representative	Architect, or Clerk of Works
Adjudicator	Engineer and Arbitrator	Arbitrator
Compensation Events	Claims and Variations	Claims and Variations
Delay Damages	Liquidated Damages	Liquidated Damages
Equipment	Plant Temporary Works	Plant and Equipment
Works Information	Specification and Drawings	Specification/ Schedule of Work

Figure 8.3: A Comparison of Contract Terminology.

The Supervisor's role is to check that the works are constructed in accordance with the contract; again the supervisor has no obligation to act impartially and this has the potential to undermine the 'spirit of mutual trust and co-operation' as he is on the Employer's payroll. Nevertheless recourse to the Adjudicator (clause 90.1), if the Contractor believes conduct is not in accordance with the contract, may counterbalance these concerns. Under clause 10, the Adjudicator is required to act in a 'spirit of independence'.

The text of the NEC is written simply and in the present tense. It does not cross-refer clauses and it makes ample use of bullet points to improve readability. The sentences are short (upto 40 words) in contrast to other popular forms of contract (e.g. there is a 241 word sentence in the ICE conditions War Clause – which thankfully has never been used in the UK).

The language has inevitably brought its critics. Some have asked how the simple English will stand up to 'loop-hole engineering', while others have seen it as a "...sacrifice of brevity in the interests of clarity". One commentator has even said that, although the NEC could be understood by someone whose mother tongue was not English, perhaps this had been at the expense of those whose it was. Whether the NEC terminology is merely a game of semantics will only be borne out in its future use.

(3)   Flexibility

The ECC Guidance Notes state:

> *"The ECC is intended:*
>
> - *to be used for engineering and construction work containing any or all of the traditional disciplines such as civil, electrical, mechanical and building work.*
> - *to be used whether the Contractor has some design responsibility, full design responsibility or no design responsibility.*
> - *to provide all the normal current options for types of contract such as competitive tender (where the Contractor is committed to his offered prices), target contracts, cost reimbursable contracts and management contracts.*
> - *to be used in the United Kingdom and in other countries."*

The above 'types of contracts' would appear to refer to alternative contracting strategies and options for reimbursement, rather than actual different types and forms. Notably there is no 'design and build' option as, under clause 21, the extent of Contractor's design input may vary to suit. In practice this means anything between 0% and 100%, provided the Works Information adequately prescribes that which has been designed and that which is to be designed. Hence the NEC provides for a great variety of uses.

Furthermore, with the combination of core and secondary (optional) clauses, the client may selectively choose which set of conditions are most appropriate, thus allocating the risks and structure of power accordingly. The more contentious issues (such as performance bonds, trust funds and delay damages) have been kept as optional clauses to be used if considered appropriate. Thus the emphasis is placed on the client to think through his contracting strategy in advance of the works.

When compared to the other standard forms of contract in the industry, the NEC's optional clauses provide a robust and flexible approach to the contingencies of contracting in a project environment. This flexibility has two major advantages:

1. The NEC provides for a variety of contractual approaches to risk apportionment depending on the contingent circumstances of each project[10].
2. The NEC provides for a variety of contractual approaches to the management of the relationship with the contractor.

These are key features of any contract which attempts to manage business in an uncertain environment, such as construction. The issues of aligning an appropriate contract with a governing relationship will be covered in greater detail in Chapter 16. The advantage of the NEC over other standard forms of contract is that it is sufficiently flexible to be employed in a number of different circumstances without having to resort to using another form of contract. However this is not the only consideration required in matching the appropriate contract with a governing supply relationship, as is discussed in the following sub-section.

(4)   Mutual trust and co-operation

The obligation on the parties to act 'in a spirit of mutual trust and co-operation' is prescribed in clause 10; the words have been lifted straight from the Latham Report. The NEC Subcontract places a similar obligation on all parties along the supply chain (clause 26.3). This requirement is perhaps the single-most important feature of the NEC, for without it much of the contractual mechanism would break down. But, as eluded to in the previous sub-section, should the contract stipulate the type of relationship, or should the relationship govern the type of contract to be used?

In clause 10 the NEC has clearly documented the type of relationship to be adopted in its transactions. However the notion of an obligatory condition to contract in trust and co-operation is curious. It is questionable how the Courts would take a view on such a clause, or indeed whether this is enforceable. Indeed the legal effect of acting 'in a spirit of' mutual trust and not simply acting with mutual trust is questionable; especially if there is dissonance between the intentions and actions of a party.

However this, perhaps, has missed the point altogether. One of the intentions of the NEC is to completely avoid the Courts, and therefore disputed matters and adversarial relations would imply the NEC has failed. The question therefore arises whether it is the contractual conditions or the contractual relationship which determines performance. It often said that successful contracts are those where there is no reliance on the contract documents (i.e. 'it is left in the drawer'). Indeed, it is not strictly necessary to have a written document at all and now there is a contract totally based on flow charts instead of clauses[11].

A balance needs to be found between these extremes; the notion that the relationship and contract are not inter-dependent is surely flawed. The terms of the contract may influence and shape the relations to produce successful performance, however one without the other is unlikely to work. In these instances the NEC has got it the wrong way round: the *relationship* needs to pre-exist and govern the behaviour of the parties in order for the contract to succeed in a spirit of mutual trust and co-operation!

Thus a *relationship* of 'mutual trust and co-operation' needs to be established between the parties in order to contract this way **and this is not addressed in the NEC**. The client needs to ensure he is able to select the appropriate suppliers and control the supply chain in this way. Appropriate supply and value chain analysis is necessary to establish fit-for-purpose supplier selection and relationship management. This theory has been developed by Andrew Cox elsewhere[12].

For the NEC this means that adversarial arms-length contractual relations are not possible, even if they are the most appropriate means of ensuring the construction works achieve their business objectives. A practical example of this is shown in Case Study 7 towards the end of this chapter. When the relationship is non-collaborative, there are insufficient oints of leverage in the NEC to deal with a non-compliant party.

Thus in order to employ the NEC successfully, a preferred supply relationship needs to be established where there is a *coincidence of interests* between the two parties in order to allow trust and co-operation to mutually flourish. In practice there are a number of ways of achieving this and this is discussed in greater detail in Part E of this book.

Thus, where a collaborative relationship pre-exists and governs the behaviour of the parties, it is advocated that the NEC will provide an excellent framework for contracting the works particulars. Ironically, however, when such a relationship exists, the need for the NEC may very well be negligible as neither party would choose to rely on its contractual force and risk breaking the relationship that exists. Thus the only conclusion that can be drawn from this is that the contractual term to act in a spirit of mutual trust and co-operation is meaningless.

(5)   Project management

One of the most significant underlying principles of the NEC is that the contract should be a management tool which sets out good management procedures[13]. The ECC Guidance Notes state:

> *"The two principles on which the ECC is based and which impact upon the objective of stimulating good management are:*
> - *foresight applied collaboratively mitigates problems and shrinks risk, and*
> - *clear division of function and responsibility helps accountability and motivates people to play their part."*

In effect the ECC achieves this through:

- active planning and programming procedures by both parties including resource planning and the effect of compensation events on timing and float within operations (ref. Clauses 31 and 32);

- early warning procedures to alert the other party to any matter affecting the price, completion date or performance (quality) of the works (ref. Clause 16);
- advance quotation and assessment procedures to determine the effect of compensation events on the price, and also to ensure the Final Account (and Contractor's cashflow) is known up-front (ref. Clauses 62, 63 and 64);
- tight response periods for administrative matters;
- quick resolution of disputes in order to minimise their incidence and severity.

These are all very commendable aspects which promote good (pro-active) contract management, which are not prescribed in other popular standard forms. They cannot however be seen as a cure for all ills, or as a substitute for good management itself. Furthermore, we suggest this approach addresses the symptoms and not the root causes of contractual problems. Loosemore[14] has suggested that one of the main issues associated with unexpected problems in the course of a contract's duration is the inappropriate behavioural responses from an organisation's members. This leads back to the previous discussions on industry culture and on the importance of *relationship*.

However, perhaps just as significant, is the question of the cost of these procedures. The Employer now has a minimum of two representatives on site to pay for, as well as the shared cost of an adjudicator when referred to. There are already indications that the administration of early warning meetings, quotation and assessment of compensation events, and a continuous revision of the programme is likely to be a burdensome (and expensive) challenge for site management. This also raises questions concerning the use of IT at site level (perhaps already mandatory on large projects, but the smaller ones too?) Inevitably the client is paying for all this, which will need to be justified in some way. The argument from the NEC camp appears to be that the increased 'up front' administrative requirement will reduce 'tail end' dispute costs. However, when asked to dip a little deeper into their pockets, some clients may need some more persuasion than this!

As with the choice of contracting strategy, the emphasis is firmly placed on the client to provide well-specified Works Information prior to the formation of the contract. Failure to achieve this (however good the subsequent project management is) is guaranteed to generate time and cost over-runs which is likely to detract from the intent of the works.

A Summary of the Merits of the NEC

The above has attempted to be an objective and thought-provoking critique of some of the key elements of the NEC. A review can become a sterile task and it would be easy to overlook the positive measures which the NEC instils. That the contract has achieved its three principal objectives is almost a certainty: it is easy to comprehend, has a wide variety of uses, and will certainly stimulate users to consider its use very carefully. Clearly many of its procedures have not been adopted in other standard forms of contract and in this respect the NEC family can stand alone.

Site administration under the contract should improve, especially with the programming requirements and tight response periods within the ECC. This may serve to improve site records and, in tandem with the early warning procedures, may reduce the incidence of claimsmanship. Similarly the re-definition of the role of the Engineer (and/or Architect) seems to offer a positive contribution to successful contract administration. The real test for the NEC will be its performance over many years with billions of pounds value of work, not just a handful of cases during the NEC's honeymoon phase.

Unlike other popular forms of contract which tend to be constructed as 'reactive mechanisms' against the incidence of certain events, the NEC is an attempt to introduce advance planning. Its proactive approach is hoped to lead to greater dispute avoidance, increased administrative efficiency and good project management. This issue of *dispute avoidance* through good project management is also re-visited in Chapter 14. The NEC clauses attempt to mould behaviour, requiring parties to enact their responsibilities rather than adopt the usual 'goalkeeper' reactive mentality to contracting and its risks. However whether this will be successful would seem dependent on the willingness of the parties to participate, rather than the legal force of clause 10.

In the absence of mutual trust and co-operation or, more specifically, a *governing relationship* of mutual trust and co-operation, the NEC is riddled with potential problems. Some of these have already been discussed and concern the management of the contract and its administrative issues, an interesting practical example of this can be seen in Case Study 7 at the end of this chapter. There are also some specific legal issues of concern:

- The ECC is untested in the Courts and there is no judicial interpretation of some of the NEC terms and phraseology.
- The title of goods clauses (70 to 73) look set to produce an administrative nightmare unless the Employer is willing to accept the risks of title on equipment, plant and materials being stored on a site under the management of the Contractor. The Employer's title on the goods is dependent on the Contractor having title, rather than the passage of consideration between the parties (the NEC does not have a provision for the reimbursement of 'materials on site'). Thus, if the Employer wishes to know whether it is carrying the risk of certain goods or equipment, it is required to determine their title of ownership from the contractor, which is not a simple administrative procedure at site level.
- As several commentators have been quick to comment, the liquidated damages (or rather 'delay damages') could become nullified if partial take-over occurs.
- The provisions for adjudication do not reflect the time period requirements of the Housing Grants Construction and Regeneration Act 1996, in that the ECC allows the adjudicator 69 days before reaching the decision, whereas the Act requires 28 days (ref. clause 108(2)). As such, this will nullify the use of the NEC Adjudicator's Contract and deem the Secretary of State's "Scheme for Construction" Part 1 to apply.

There are other notable difficulties. These include:

- the difficulty of preparing and assessing quotations for compensation events in both terms of time, cost and administration. This could be particularly onerous for small compensation events where the cost of administering the transaction could exceed the value of the event, or when there are several [separate] compensation events coinciding at the same time. (This particular point was an issue when the National Rivers Authority trialled its first use of the NEC in 1994 on the Lyme Brook Flood Alleviation Scheme. The burden of the NEC's procedures meant that the Project Manager became heavily embroiled in the administration of the compensation events at the expense of his other contractual duties).
- The possible need to reorganise internal structures and the provision of the appropriate level of delegation to the Project Manager. This will also lead to the requirement for appropriate training and education of staff.
- There are also some notable omissions in the ECC (such as dayworks, provisional sums, PC items, nominated sub-contractors, provision for materials on site, collateral warranties, etc.). It is arguable that, as some of these are sources of significant problems in other contracts, it is therefore better that these are omitted altogether. However not all will subscribe to this view.

Structure of power

The advent of the NEC has provided tangible evidence of the power struggle within the construction industry's supply-base. This is typified in the way in which the NEC removes the client from the day-to-day proceedings and relinquishes its control to the suppliers. The client takes no part in the progress of the works other than for reasons of payment. This therefore creates dependency in two areas: (1) the adequacy of the contract's front-end specification and (2) the competence of the Project Manager and Supervisor to administer the contract. In either case the client is dependent on its professional advisers.

The ECC has been crafted to instil specific project management procedures to the works process. Nevertheless, this is still at the client's expense; the contract merely provides a framework for the contractor and the third party professionals to work together; there is no direct benefit for the client. Although an immediate response might suggest this is in the client's favour, the obvious retort is that the client's suppliers have already been contracted to deliver the works and should not need to have formalised administration procedures super-imposed on their work activities in order to improve their competence.

The difference between the NEC's structure of power and other standard forms of contract is merely the existence of a more regimented procedure of 'good practice'. It not only supports the position of the industry's professional suppliers, but also creates another cost-structure to present to the client. The question that therefore follows demands whether all this is really necessary.

If, as this chapter has suggested, the presence of a governing relationship of mutual trust and co-operation already exists between the client and contractor, then why is there the need to formalise a cost-adding structure of procedures to the commercial proceedings? There is little evidence that this is actually adding any value to the client; rather, it is simply adding cost to the transactions and justifying the employment of yet more professional advisers. In short the structure of the NEC supports the power-base of the industry's professions, at the expense of its clientele.

When should the NEC be used?

The key question that arises from this analysis is under what conditions can the use of the NEC be recommended?

In 1997 the Construction Clients' Forum produced a leaflet which comprised a 5-step 'independent' guide to value for money for the lay-client, of which Step 2 (The Project Strategy) endorses the use of the NEC for all forms of construction[15]. Interestingly the leaflet was prepared by a small consultancy firm which was also involved in advising the NEC Working Group and the Latham Review. Is this 'independent' advice appropriate? The answer is almost categorically not. ***There are clear circumstances when the NEC should not be adopted.*** Even large experienced clients are still coming to terms with this new contract (see Case Study 7). How much more so will small inexperienced clients?

Case Study 7: Problems of Implementing the NEC

Several UK organisations have been involved in a series of trials of the NEC to test whether its novel features are suitable for their purposes or not. One such client, with an annual construction spend of £180 million, made its findings known to the authors, but because of its public connections with the NEC preferred to remain anonymous. These experiences are based on the use of the ECC during 1996 on a range of construction projects up to £7 million in value and have been recorded below for the reader's benefit:

- The NEC necessitates clients to be precise and detailed in the Works Information to prevent burdensome administration in the post-award phase.
- The NEC programme needs greater evaluation than other programmes during the tender stages.
- There is no action in the event of the contractor not updating the programme, thus making compensation events very difficult to analyse.
- Problems have been encountered assessing the 'actual cost' associated with compensation events: they do not refer to tendered bill of quantities rates and the employer has had difficulties in accessing subcontractor prices. The result is that the employer has very little to assess prices against.
- Problems were encountered when Project Managers were expected by the ECC to act on issues that were beyond their corporate level of delegation.
- The organisations' staff experienced difficulties due to lack of familiarity and support despite attending pre-contract training workshops.
- When under pressure by the ECC's timescales, there was a tendency for the parties to resort to traditional adversarial behaviour.
- Because of the non-confrontational approach required by the NEC, there remains little leverage to use against a non-complying contractor.
- Because of procedural requirements, additional site presence is required.
- The indication of this review was that the NEC should not be used on contracts less than £1 million in value or for less than 4 months in duration.

Furthermore it is arguable that most small and occasional clients will be buying from the construction industry on a one-off transactional basis. Such purchasing conditions often dictate multiple sourcing and competitive tender in order to gain maximum leverage on the transaction (unless the construction works are of exceptionally strategic importance to the client firm)[16]. These conditions will almost always induce arms-length adversarial relations. The NEC is, arguably, inappropriate for these conditions and cannot be recommended.

Conclusion

The NEC is a brave attempt to find another way of contracting in the construction industry by formalising management procedures and determining a governing relationship. Whether it succeeds look set to depend on the volition of the parties to override the existing cultural and structural barriers within the industry. Whether or not one should adopt the NEC is still open to debate. Certainly there are some good 'ideals' embedded in its mechanism, but whether they can be successfully implemented in today's industry is yet to be proven.

This chapter has examined the NEC's structure of power and whether it will succeed in today's construction industry. Certain reservations have been expressed concerning its ability to shape contractual relations in a spirit of mutual trust and co-operation (and, even, whether this is desirable) or whether the clause is, in practice, meaningless. Given certain pre-requisites, the NEC has a contribution to make to the construction industry and may serve to be fit-for-purpose in particular pre-determined collaborative circumstances. However, on its own, it still fails to address how the client or contractor should select and create supply relationships to deliver successful construction works. Without the presence of a governing close-working collaborative relationship between buyer and supplier, the NEC is set to fail.

Chapter Notes

1. Perry J. (1995) 'The New Engineering Contract: principles of design and risk allocation' *Engineering, Construction and Architectural Management* Vol. 2, No. 3, pp. 197-208.

2. There are many commentaries and articles on the introduction of the NEC. Many of its protagonists are associated with the NEC Panel and some of their publications show remarkable similarities. Refer to Barnes M. (1994) 'The New Engineering Contract: a promising start' *Proceedings of the ICE* Vol. 102, Iss. 3 (August), pp. 94-95; Broom J. & Perry J. (1995) 'Experience of the use of the New Engineering Contract' *Engineering, Construction and Architectural Management* Vol. 2, No. 4, pp. 271-287; Barnes M (1996) 'The New Engineering Contract – An Update' *International Construction Law Review* Vol. 13 (January), pp. 89-96; or Baird A. (1995) 'Pioneering the NEC system of documents' *Engineering, Construction and Architectural Management* Vol. 2, No. 4, pp. 249-270.

3. Refer to Broom J. (1997) 'Best Practice with the New Engineering Contract' *Proceedings of the ICE* Vol. 120, Iss. 2 (May/June), pp. 74-81.

4. There have numerous suggestions, including an excellent study by P. Ireland (1995) *The influence of the communication flow requirements of the NEC on the efficiency of the contract and organisation concerned*, unpublished Masters dissertation, University of Birmingham.

5. Latham M. (1994) *Constructing the Team - Final Report of the Government/Industry Review of procurement and contractual arrangements in the UK construction industry*, HMSO, London.

6. For an interesting discussion of the effect of the NEC on the industry's prevailing culture refer to: Rooke J. & Seymour D. (1995) 'The NEC and the culture of the industry: some early findings regarding possible sources of resistance to change' *Engineering, Construction and Architectural Management* Vol. 2, No. 4, pp. 287-306.

7. Refer to: Cox A. & M. Townsend (1998) *Strategic Procurement in Construction*, Thomas Telford, London.

8. Refer to Baird A. (1994) 'The New Engineering Contract – A Management Summary for Plant Industry Users' *The International Construction Law Review*, Vol. 11, Part 2, pp. 114 - 127; and Baird A. (1995) 'Pioneering the NEC system of documents' *Engineering Construction and Architectural Management*, Vol. 2, No. 4, pp. 249 - 269.

9. Perry J. (1995) 'Structuring Contracts for the Achievement of Effective Management' in J. Uff and M. Odams (eds.) *Risk, Management and Procurement in Construction*, Kings College London, London, pp. 357-374.

10. It is imperative that any risk assessment first identifies the firm's business goals at a strategic level and any risk of failing to achieve those goals and then proceeds to manage those risks commercially so as to minimise the risk exposure of the firm at a strategic level. For a useful guide on strategic risk assessment in construction refer to: the Construction Round Table Guide to Strategic Risk Auditing prepared by Andrew Cox Associates.

11. Refer to: Abrahamson M. (1995) 'Risk, Management and Procurement and CCS' in J. Uff and M. Odams (eds.) *Risk, Management and Procurement in Construction*, Kings College London, London, pp. 375-392.

12. Refer to: Cox A. (1997) *Business Success: A Way of Thinking About Strategy, Critical Supply Chain Assets and Operational Best Practice,* Earlsgate Press, Boston, UK.
13. Perry J. (1995) 'The New Engineering Contract: principles of design and risk allocation' *Engineering, Construction and Architectural Management* Vol. 2, No. 3, pp. 197-208.
14. Loosemore M. (1994) 'Dealing with unexpected problems - do contracts help? A comparison of the NEC and JCT 80 forms' *Engineering, Construction and Architectural Management* Vol. 1, No. 2.
15. Construction Clients' Forum (1997) *Thinking of Building? A 5 Step Guide to Value for Money.*
16. For further discussion refer to: Cox A. & M. Townsend (1998) *Strategic Procurement in Construction*, Thomas Telford, London.

Learning Resources
Centre

Chapter 9

Other Forms of Contract

Introduction

In practice, most decisions in construction procurement have been limited to the choice between the sequential contracting method and that of design and build. These options and their supporting forms of contract have been explored in some detail in the preceding chapters. It is important, however, to recognise that these represent only the two most popular contracting methods, of which there are several. In Chapter 4 these other methods were briefly outlined and included: performance contracting, management contracting, framework agreements, extended arm contracting and relational contracting strategies. These other methods should be given equal consideration in determining the optimal procurement route.

The preceding chapters have examined various contracting methods and their respective forms of contract, with an emphasis placed on the structure of power which the contract gives the transaction. This chapter continues this theme by considering three further contracting methods (construction management, management contracting and performance contracting). Not all of these methods have industry standard forms of contract and this chapter limits its examination to the JCT MC/87 (for management contracting) and the IChemE Green Book (for cost-reimbursable performance contracting).

Management Approaches

Management approaches to construction have been gradually gaining popularity in the industry over the past 20 years; indeed some even consider their popularity will continue to increase to as much as 15% of the market by the turn of the millennium[1]. Generally there are two principal management approaches:

- *Construction Management* (CM) – where the client employs a professional management team to co-ordinate the work packages and to form a series of directly employed works contracts between the client and trades contractors;
- *Management Contracting* – where the client employs a contractor to co-ordinate the work packages and to subsequently enter into works contracts with the trade contractors.

Construction management

CM lost popularity in the UK during the late 1980s and the early 1990s due to the repeated number of contractual and commercial problems it generated. One well-known case, the new British Library at St Pancras in London (see Case Study 8), has been endemic of the problems that can occur with this contracting method if managed inappropriately. The project was subject to vast cost and time over-runs and, as such, has been under the spotlight of the press for a number of years. As a result, the press became critical, not just of the way in which this particular project was managed, but more generally of the CM method as a whole.

In general CM is used for large and complex projects of a once-off nature. It is particularly appropriate where there is likely to be a number of specialist trade suppliers working on the project at the same time. These might include electrical contractors, fitters, quantity surveyors, steel erectors, service engineers and plant installation contractors, *inter alia*.

The advantages and disadvantages of CM are associated with the *structural* provisions of the method illustrated in Figure 9.2. As a rule, CM requires greater input, co-ordination and management from the client and, in theory, this investment is repaid by the fact that the suppliers are not required to manage the project as a whole. This shifts the balance of power away from the traditional approaches to contracting by firmly keeping the responsibility and power to act with the client. This should bring a cost saving to the client, although in practice it is usually set-off against the added transaction costs associated with CM. In general the works are fragmented into areas of specialism which, generally speaking, means the works contractors are smaller and carry less corporate overheads. However, it also means that there is a lot of additional cost involved with administering the individual contracts.

Thus, with CM, the client has considerable exposure to the burdens of risk and reward potential. If the client can manage (control) the process competently, then it follows that significant benefits can be realised. However, it also follows that the client suffers the full risks of project mis-management, if it loses control on the process, as in the case of the new British Library.

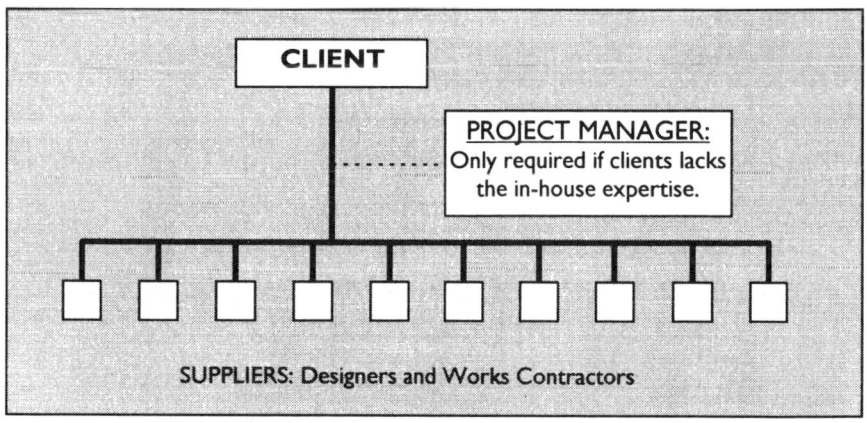

Figure 9.2: The Structure of Construction Management.

The essence of CM is good project management and the client contracts directly and separately with each supplier (see Figure 9.2). Flexibility is key and the client needs to retain absolute control on the contracting process as each contract is tied into the delivery of the completed project. Should any one individual contract over-run or suffer unmitigating events which cause it delay, there will be a consequential knock-on effect for the remaining contracts.

Thus the client needs to manage the process tightly in order to control the cost of the project. Each contract needs to be managed discretely from the others and in such a way that allows maximum flexibility. For the client who is able to do this competently, the CM approach is highly appropriate; the client retains the full potential for leverage over the supply-base.

At present there is no standard form of contract for CM; each agreement is subject to the individual needs of each works package. That is not to say, however, that some of the standard forms of contract could not be used for some of the works packages (e.g. the ICE 6th for ground works, the JCT IFC'84 for building works and the NEC Plant Contract for installation of plant). The key is to employ the contract which is most appropriate and, for this, some clients will draft their own bespoke contracts. Despite adding to the transaction costs, this allows the client to select the most appropriate conditions for the specific works in mind. As most CM projects are of high value, this additional expense is likely to be worth it.

Key elements of the individual works contracts should include the abilities to expedite the works (should progress become delayed) and to control the 'window of access' to the site, in terms of mobilisation on and off site. Where the client does *not* have the leverage to control the works contractors, additional contractual incentives may also need to be built into the contract to ensure delivery to time and budget. This helps to generate a *community of interests* between client and supplier, thus helping to overcome supply-base opportunism that comes with one-off projects.

Thus there is a pressing need for clients wishing to pursue the CM approach to ensure they establish the optimum contractual relationship with their suppliers. Individual contracts need to have a structure of power that enables the client to manage the whole project and not just the works packages in isolation. There are considerable advantages to be realised from the method if it is managed well and transaction costs are kept to a minimum. Similarly there are considerable disadvantages to this method if it is managed incompetently leaving the client vulnerable to considerable risk.

Case Study 8 demonstrates a clear example of a poorly managed CM project which resulted in major cost and time over-runs. The reasons for this lack of competence are transparent. Firstly, the split-responsibilities of two government departments meant that there was no single decision making body operating as the client. CM requires strong hands-on project management and the ability to direct and make swift concise decisions as the project progresses; neither government department was able to do this and furthermore, together, they inhibited each other. Indeed, 'dual' client roles should never be matched with a fragmented supply base; an alternative method to CM should have been adopted under these conditions. Secondly, the works contracts were considered by the National Audit Office to be inappropriate in terms of roles, responsibilities, authority and conflicting objectives[2]. The result was that, irrespective of the quality of the on-site management and construction capabilities, the works were always going to fail because of the structural properties of the project. Clearly, CM was an inappropriate contracting method for these contingent circumstances and this was further exacerbated by incompetent management in both government departments throughout the 1980s.

Case Study 8: The New British Library

The aspiration for a New British Library at St Pancras, London was born out of the British Library Act 1972 which amalgamated a diverse number of institutions spread across some 20 buildings in London. The original plans were for a 200,000 square metre building, but this was later reduced to an area of 108,000 square metres with 335 linear kilometres of shelf space.

The project was originally commissioned by the Property Services Agency (PSA) on behalf of the Office of Arts and Libraries. Construction had been planned to commence in March 1978 but, due to intense pressure on public spending, the start was delayed until April 1982, when the decision was made to procure the building through the construction management route. At the time, no limit on spending was established; the project was to be financed by year-on-year annual funding.

By 1988, it was clear that the project was not being managed well and remedial action was required. The financial responsibility for the project was passed to the Office of Arts and Libraries, while the PSA retained responsibility for the project's overall management. At that time Ministers set their first limits on the project of £300 million with completion programmed for 1993.

In the course of 1990, the project was re-examined and a further £150 million were allocated to its completion. In October 1990 the National Audit Office (NAO) reported progress on the project and concluded that its management lacked clear definition. It stated that uncertainties about the design and funding of the building had extended the timetable considerably. In August 1991, the Government declared that they expected the building to be completed in 1996 within the new cash limit of £450 million.

As the project progressed further time delays and cost increases were being accrued. In the course of 1994, TBV Consult advised that these delays were costing approximately £1.5 million a month and extensive negotiations were conducted with the project's key works contractors to establish milestone payments for delivery. In November 1994, Ministers agreed a further budget increase on the project to £496 million.

By this time, the project had gained increasing attention from the public media who were critical of the way in which the government was handling the project. There was increasing pressure on the project management staff to ensure that no further delays or cost increases were incurred.

These pressures were intensified in 1996 when the NAO published its second report and slammed the way in which this project's procurement had been managed. In particular they reported:

- deficiencies in defining roles and responsibilities prevented the project strategy from working effectively and contributed to the difficulties in controlling the project;
- there was a complex network of contracts in which the most important did not fit together well (the contracts did not reflect either current or past best practices and, as such, contributed to confusion and conflict);
- financial control was not being maintained since budget increases were being approved retrospectively to reflect costs already incurred by delegated authorities;
- 26% of the project's prime costs related to professional fees (£100 million) of which most were earned on cost-plus arrangements which did not give any incentives to ensure progress to time or cost.

These are alarming conclusions drawn by the auditors which expose considerable mal-practice and incompetence throughout the management of the project. Since the NAO's second report, it is believed that claims for additional expense have continued. At the time of publication, the project is still not complete and the total costs of the project are £511 million.

Source: NAO[2]

Management contracting

At first glance, management contracting could easily be confused with a traditional approach to contracting with 100% of the works sub-contracted out to others. Indeed many UK contracting firms predominantly operate in a management capacity with the works being carried out by their sub-contractors. Dearle and Henderson state: *"The essence of [management contracting] is that it seeks to separate the responsibility for management of available resources from the actual construction, by removing or reducing the risk element of the firm providing the management skills."*[3] Thus the Management Contractor only carries out a management function while the works contractors perform the actual construction operations. In other words it is similar to the CM method except that the client has outsourced the management process of the works (see Figure 9.3). These works operations might be packaged into: groundworks and foundations, structural services, steel erection, general building works, mechanical works, electrical works, services and facilities management, *inter alia*.

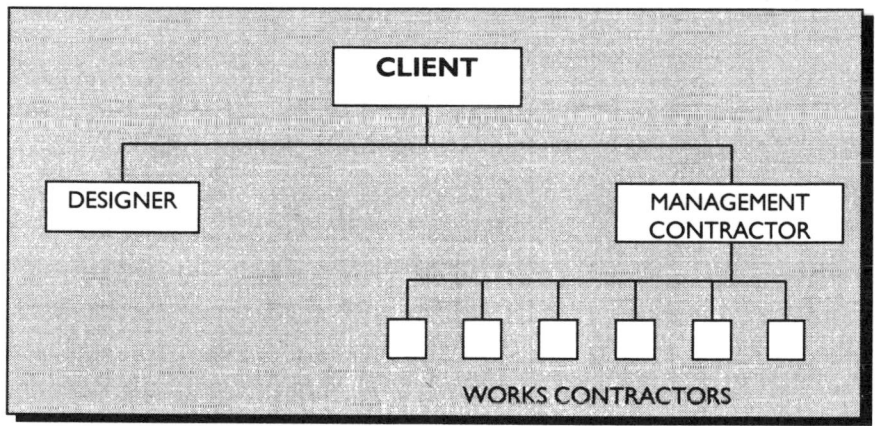

Figure 9.3: The Management Contract Structure.

The Management Contractor exercises co-ordination, time, cost and quality control over the work package contractors and provides facilities for their common use[4]. Typically the Management Contractor's tasks might include:

• programming;

- resource allocation;
- logistics;
- packaging of works;
- contract administration of works contracts;
- liaison with professional team (i.e. architect and quantity surveyor);
- site management and quality control.

ADVANTAGES	DISADVANTAGES
1. Overlap of design and construction leading to early start on construction.	1. Client may be exposed to greater risk from construction contractors.
2. Potentially quicker completion.	2. Uncertainty about the cost of the complete works at the start of construction.
3. Involvement of management contractor at the planning stage could lead to better packaging of the works.	3. Doubts about liabilities of each party can cause difficulties.
4. More flexibility during construction.	4. Roles and responsibilities of designer and Management Contractor for quality control are unclear.
5. More help when there are problems.	5. Tendency for additional administration and some duplication of supervisory staff.
6. Reduced claims from knock-on effect.	6. An extra (unnecessary) tier of management (site contractors also have site management).
7. Simpler lines of responsibility.	7. Management Contractor's liability often limited to professional negligence.
8. Advice of Management Contractor at design stage leads to an early input of buildability advice.	8. Letting works in phases means that professionals (e.g. surveyor) must be more involved.
9. Increased reliability of cost and time estimates.	9. More paperwork and transaction costs involved in bid packages.
10. More realistic planning.	10. Potentially lavish expenditure by the Management Contractor on site establishment, plant and other preliminaries.
11. Fewer claims.	11. Less control for the client on the process.

Figure 9.4: The Merits of Management Contracting.

Source: adapted from Ward et al.[5]

One commentary[5] considers that the merits of management contracting can be summarised under the four headings of: fast completion, improved design, minimised costs and better supervision and co-ordination. Rather inconclusively, it examines these merits and concludes that: (1) quicker completion could result from improved planning and overlapping activities (as with any project planning under any other contract); (2) the design produced under a management contract is not significantly better (in terms of ease and/or expense to build); (3) there is no evidence of reduced construction costs; and (4) that the claims of better supervision and co-ordination are countered with issues associated with unclear roles, responsibilities and liabilities of each party and a tendency for additional (unnecessary) administration.

The conclusions are somewhat mixed about the merits of management contracting, as shown in Figure 9.4. Furthermore, in the following examination of the JCT MC/87 form of management contract, it is demonstrated that the client has little or no control (power) to control time or cost throughout the project.

The JCT Management Contract MC/87

Until the publication of the New Engineering Contract in 1991, the JCT Management Contract (MC/87) was the only standard form for management contracting in the construction industry. As well as straight management contracting, this contract can be used when the Management Contractor is made responsible for the design (the *Design and Manage* method), or where the Management Contractor assumes some direct responsibility for the performance of the trades contractors (the *Design, Manage and Construct* method). Either of these subsidiary approaches to management contracting can be procured through the JCT MC/87 by the simple amendment of some of its clauses.

A full list of the Management Contractor's responsibilities are listed in the Third Schedule of the MC/87. These are generally divided into two categories: those services to be supplied prior to construction and those to be supplied during the construction period. Relatedly, the Management Contractor's fees are also split between the Pre-Construction Period Management Fee (PCPMF) and the Construction Period Management Fee (CPMF), with the latter being subject to 3% retention by the Employer until the Final Certificate is issued (clause 4.12).

Notwithstanding this, it must be emphasised that the MC/87 is not a lump sum contract. In fact *the client does not know how much the works will actually cost until they are complete.* As an indication of the price the client will pay (exclusive of the Management Fee) a Contract Cost Plan is prepared in advance of the works by the Quantity Surveyor based on the project documentation (drawings and specification). The actual cost incurred by the Management Contractor and the works contractors is the amount payable by the client; this actual cost is called the Prime Cost. Over and above this, the Management Contractor's fee comprises either a pre-agreed lump sum or a pre-agreed percentage of the Contract Cost Plan Total. Either fee arrangement is effectively a fixed fee, unless the Contract Cost Plan is changed. It should be noted that the NEC Management Contract differs slightly in regard to this, as the management fee is based on a percentage of the actual cost (i.e. a cost-plus arrangement), and thus will increase if the works contract sums increase[6]. The total amount payable by the client under the MC/87 can be illustrated as shown in Figure 9.5.

Thus with the MC/87, the Management Contractor's fees are fixed and do not fluctuate with the prime cost of the works. Nevertheless, some have argued that management contracting can be more expensive than other forms of procurement because the Management Contractor has no incentive to keep the costs down[7]. Some empirical evidence has suggested management contracting to be more profitable for the Management Contractor than traditional methods of contracting[8]. Similarly, Sidwell claims that a management contractor's expenditure on site establishment, offices, plant and other preliminaries are generally greater on management contracts[9]. Logically, this stands to reason as the Management Contractor is reimbursed all of his costs (regardless of the sum) but earns only a fixed fee; therefore the Management Contractor will seek to have as much reimbursed as 'cost' (i.e. on-site expenses rather than head office overheads) in order to leave the 'fee' free as a profit margin. In this way the JCT MC/87 is demonstrably drafted against the client.

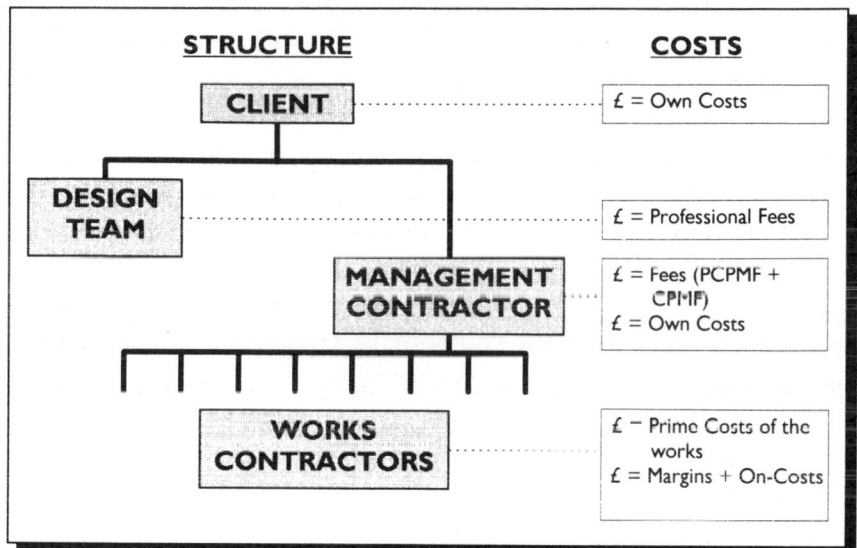

Figure 9.5: The Costs of Management Contracting.

Furthermore, *the Management Contractor has no incentive to minimise the project costs*. There is no incentive to control the costs of the works contractors' contracts or to exercise financial controls on variations and/or claims. Indeed the aforementioned Dearle and Henderson quote regarding the segregation of risk and responsibility is more fully understood in the light of this; *the Management Contractor has little or no responsibility for the costs of the works.* The risk of increased costs has been removed from him and is carried by the client and/or works contractor or, in times of negligence, the works' designers. Thus the Management Contractor enjoys responsibility without the risk of the consequences (except in negligence); while the client bears the risk without any management control. For most clients this is totally unacceptable and remains the antithesis of good practice.

In short, the balance of power is heavily tipped against the client; to procure construction through the JCT MC/87, the client is required to hand out a blank cheque-book to the supply base. Never has the incentive to make money at the customer's expense been so crudely expressed!

However there are some notable advantages of the management contract (as already stated) for which certain construction projects will be well-suited. The general consensus of opinion from the legal commentators is that the JCT MC/87 is appropriate:

1. when projects are large and complex;
2. when there is a need for flexibility;
3. when there is a need for early completion.

One essential proviso to the above concerns the employer's desire to change his requirements in the course of the contract's duration (clause 3.4). Since the employer has devolved responsibility through the Management Contractor to a complex supply-base of interdependent works packages, changed requirements could result in heavy consequential effects on the works in terms of loss and disruption. As Murdoch and Hughes state, changes should be *avoided at all costs*[10].

Summary of JCT MC/87

In summary, the MC/87 provides the construction industry with a management contract suitable for the procurement of large multi-disciplinary projects, where time to completion is important and the client is looking for a party to co-ordinate all activities to deliver the completed works. One particular advantage is that the Management Contractor can package and resource the works, utilising his own expertise, by overlapping activities and thus expedite the completion date.

However, from the client's point of view, there have been several disadvantages with the JCT MC/87 highlighted in this section, including:

- the lack of incentive to minimise costs to the client (although in this respect the MC/87 is considerably better than the NEC Management Contract);
- the lack of clarity in roles and responsibilities between the management contractor and the professional team;
- the limited liability and risk placed on the Management Contractor;
- increased (and possibly unnecessary) administration;
- the detrimental effect of changes in employer's requirements.

It has been demonstrated that the MC/87 separates responsibility from risk at the expense of the client. In this way the contract institutes a structure of power which is contrary to the client's interests and designed to compromise its business objectives.

Performance Contracting

Performance contracts are the innovation of the Institution of Chemical Engineers (IChemE). In 1964 the IChemE recognised the need for a construction contract to suit the specific requirements of its particular industry. Typically this industry (chemical and process engineering) involves the construction, operation and maintenance of complex, high-value processing/manufacturing plants. By 'plant' the IChemE means works which process one material into another on a continuous or batch basis, using either chemical, biological or mechanical means[11]. Unless specifically on a 'green-field' site, this involves interfacing with existing plant which must be kept operational in the course of the construction process. In this respect the intended use is not too dissimilar from other operationally-live project environments such as the railways industry.

The IChemE was not satisfied with the provisions of the ICE or JCT forms of contract in that they are principally concerned with ensuring the contractor has built something in accordance with a prescribed design. Plant works generally have a specific performance required of them which is unlike infrastructure, or the accommodation needs of buildings; similarly they are not expected to have anything other than a functional value. Thus the IChemE was not looking for a contract to control the construction process, but rather a contract which would enable the contractor to deliver works with specific *performance* requirements. From this sense of dissatisfaction with the existing construction contracts, the IChemE developed its own contracts based on performance and, in 1968, it published the IChemE Red Book (refer to Chapter 6).

The IChemE performance contracts are, in a loose sense, 'turnkey' contracts requiring the contractor to be responsible for the design, construction and testing of a complete operating plant. They are also 'multi-disciplinary', in that they are not prejudiced to any one industry, profession or skill-set. The contractor is expected to use whatever skills and competencies that are necessary to deliver the works to the client. As Wright[11] states: *"...the industry is mainly concerned, not with the Plant as a thing, but with the Plant as a means to provide a product. Therefore the industry uses contracts which concentrate upon whether the Plant will make the right product, rather than what the Plant will be."*

The difference in emphasis is subtle but fundamentally important. It recognises that the works are to be constructed for a **purpose** which will have a specific function to fulfil; the focus is, therefore, on the *end* rather than the *means* by which they are constructed. Thus the design responsibilities of both IChemE performance contracts are a 'fit-for-purpose' requirement and nothing else. As Wright goes on to state: *"...If the Plant does not produce the desired result....the Contractor will be in breach of the Contract"*[11].

Principally the mechanisms of the IChemE contracts are little different from other construction contracts. Generally the contractor is responsible for the same processes as any other design-build contract, but has added responsibilities for installation, running and performance testing and is required to provide performance guarantees. Rather than small individual pieces of plant and machinery being supplied and installed in isolation to the remaining works, as in other types of construction, the contracts provide for integrated construction of sizeable process plant often requiring a variety of different machines and equipment, together with their metering and control systems and support structures to be assembled within an operational production unit. As a result, much of the construction work may have been conducted off-site (particularly the mechanical and electrical works) leaving the final assembly, installation and testing to be done on site.

As a result the emphasis in the IChemE contracts (as already stated) is upon performance and this is indicated in the contracts by their specific requirements for testing, commissioning and handover of the completed works. The performance maybe in terms of product yields and efficiencies (or their equivalent), utility consumptions or any other appropriate chemical engineering criterion, as well as the mechanical soundness of the plant. It should be noted that, unless specifically stated, this does not include structural integrity in the same understanding that the construction industry gives to the design-life of new works.

It is also important to note that at the end of the defects liability period, providing the works have satisfied their required performance standards, the responsibility for the plant (operation, care, safety, servicing, maintenance, etc.) becomes that of the purchasing client[12]. The only continuing liability that the contractor owes is for the performance guarantees that have been made. This is another significant difference between the IChemE performance contracts and other contracting methods.

Both IChemE performance contracts have complementary Schedules which accompany the contract. The Purchaser is required to complete these Schedules in such a manner as to fully specify the purpose, nature and extent of performance required of the works. The extent of detail provided in these Schedules, should reflect the extent in which the client wishes to determine the completed works; insufficient detail or precision in doing this will result in unsatisfactory finished works and will contradict the 'fit-for-purpose' essence of the performance contracts. Conversely, however, the contractor cannot be expected to be liable for the performance of over-specified designs.

As a comparison, the IEE/IMechE Conditions (particularly the MF/1 Conditions[13]) are closer in philosophy to the IChemE contracts than other standard forms of construction contract. However the IEE/IMechE contracts are more specifically 'equipment-related' rather than 'plant-related' and, in general, the IChemE contracts are not suitable for the procurement of individual items of equipment on a supply-only or supply-install basis.

The IChemE Green Book

Although there is generally very little available literature providing commentary on the IChemE contracts, the Green Book seems to have attracted much discussion around the industry. Perhaps this is because of the fact it is a cost-reimbursable contract and, generally in clients' aversion to risk exposure and cost vulnerability, has consequently invited industry comment. Despite this, it remains a popular form of contract for some clients (such as Thames Water Utilities Ltd.).

Many of the performance principles of the IChemE Red Book are the same in the Green Book. The Purchaser specifies the performance requirements intended of the new plant works in the Schedules and the Contractor provides all that is necessary to design, construct and commission those works. The Green Book, by the nature of its reimbursement mechanism and some of its other provisions, is a particularly flexible contract which requires both parties to collaborate together to produce the works. It is best suited for conditions where there is a higher degree of uncertainty associated with the works at the time of contract[14], in that it allows for easy response to changes and attributes the straight forward recovery of costs with the work activities. As such the Green Book could be used for conditions, like tunnelling works (such as the successful Thames Water Ring Main project[15]), where the result to be achieved is reasonably clear but the work necessary to achieve that result is not.

Because of the presence of uncertainty, as well as the fact that the Contractor will be reimbursed for every activity it is asked to be involved in, the Purchaser (or more specifically its Agent) is required to take a far greater hands-on approach than in other construction contracts. Some have suggested that the use of cost-reimbursable contracts encourage "...an identity of interest, openness and trust..." between the parties[16]. The IChemE's Purple Book suggests that relations should not be created at 'arms-length' and thus the Project Manager has been given many more powers to control the day-to-day operations of the Contractor. In this sense, the Project Manager should only be acting as the Purchaser's Agent; the emphasis is upon control of the operations and the ability to adapt to the requirements of the works, rather than controlling the Contractor (the parties are supposed to be working together).

Therefore the contract contains considerable flexibility in that the Project Manager may instruct the Contractor to execute any part of the works at any time thought necessary (clause 11.1), or instruct the Contractor to suspend any operations at any time (clause 14.3). For such interferences, the Contractor is reimbursed any additional expense or loss suffered as a result (clause 14.4).

Furthermore, because the Contractor has been awarded a contract which entitles him to full cost-recovery for his operations, the Project Manager has the right of approval of all sub-contracts prior to appointment (clause 8.3). Thus the framework for cost-reimbursement is set. The Contractor receives full cost recovery plus a margin, while the Purchaser maintains a degree of control on what is and is not required as the works progress.

There are many perceived problems with cost-reimbursable contracts which are mainly associated with the client's liability for uncontrolled exposure to cost. The Purchaser is carrying all the financial risk and will have very little certainty of keeping to his budgeted target for the works. Conversely, for the Contractor the risk is extremely low and thus should be matched with comparatively low-profit margins. However, in practice this does not seem to carry through to the 'bottom-line' and invariably cost-reimbursement for the Contractor equates to 'cost-drain' for the Purchaser, as the overall cost of the works can be considerably higher than works of similar nature using other contracting methods. The Contractor does not have the incentive to select the materials and suppliers that offer the most value for money and is certainly not encouraged to search for cost-saving alternatives in the design/construction methods.

To overcome this, there are a number of recommended measures for improved cost control, including the application of target costing, share formulae, time targets, 'sliding scales' or guaranteed maximum pricing. For a further discussion on this reference should be made to the CIRIA report on Target and Cost-Reimbursable Contracts[16]. It should however be noted that the IChemE Green Book accounts for the time and quality targeting recommended in this report through the use of the Performance Schedules. The reader may also wish to refer to Case Study 9, which contains the published Thames Water Target Cost Formulae specifically used for the IChemE Green Book (1st Edition) and was developed by them as a public body to institute financial control on these types of contracts.

As with the Red Book, the Purchaser is required to specify his performance needs up front, but there is slightly less emphasis on the precision required. As with the Red Book, there is scope for change provided through variation orders (clause 17) and this provision is expected to be used. In repetition of what has already been stated, the Green Book is designed to be flexible to the changing needs of the works. Variations are defined as any alteration in the specified plant, or the type/extent of services provided by the contractor (clause 17.1). The costs of the variation (and those incurred in responding to it) are reimbursed to the contractor as they are implemented. In theory there is no change to the Contract Price, as the Contract Price is only formed as the works are conducted (it is the total of sums payable rather than those to be paid, see clause 38.1). Reimbursement of the Contract Price occurs on a monthly basis (clause 39.1) when the contractor submits a bill for the previous month's work plus an estimate for the next month's work. The purpose of the estimate is purely to assist with the Purchaser's budgeting, and does not serve as any fixation of price for the future.

As with the Red Book, interest is payable on overdue payments (clause 39.4) as it is a cost to be reimbursed. Similarly (and unlike the Red Book) inflation and/or price fluctuations also remain at the Purchaser's risk, although there is no specific clause to cover them. For example, although a contract may be awarded in (say) September 1996, works carried out in June 1997 are charged at June 1997 prices and not fixed at the time of award. Thus in times of inflationary markets the IChemE Green Book may be less attractive to clients.

Other specific points worth noting about the IChemE Green Book include:

- As mentioned in this report's section on the Red Book, in the Green Book there is no expectation placed on the Project Manager to exercise 'impartial judgement' (see Guide Note G). Instead he is expected to use 'the best of his skill and judgement as a professional engineer' (clause 10.1). This in effect is a 'reasonable skill and care' requirement, but it begs the question whether the IChemE expects the Project Manager to be partial in his judgements.

- The Green Book does not make any reference to the Construction (Design and Management) Regulations 1994 (which is not surprising as it was published in 1992), however, the whole contract is currently under review and thus, in this, is likely to take into consideration recent legislation.

- Schedule 8 suggests that the Purchaser may decide on an arbitrary percentage figure of the Contract Price to be inserted as the value for liquidated damages. As with the Red Book, where such a figure genuinely represents the Purchaser's pre-estimate of the likely loss in damages resulting from delay, then this practice can be considered sufficient. However, in certain circumstances, such practice could be considered to be the levy of 'penalties' rather than liquidated damages, which would result in the damages becoming unrecoverable. The Purchaser is advised to take specific legal advice before inserting anything other than a genuinely pre-estimated figure of specific value.

- The Green Book refers to *force majeure* events in clause 15. These are defined as circumstances beyond the reasonable control of either party which prevent or impede the due performance of the contract. Although in English Law there is no statutory legal concept of *force majeure*, the IChemE has circumnavigated this by allocating a specific definition to the term and listing those events which constitute it (clause 15.1). This is acceptable practice, and as such would be supported by the courts.

- Under the IChemE Green Book, there is no allowance for any form of Alternative Dispute Resolution, the parties elect either referral to an Expert (clause 44.1) for a decision or arbitration (clause 45.1). Although this could be seen as conflicting with the Housing Grants Construction and Regeneration Act 1996, this will depend on the nature of construction work involved (under the Act's definition of 'construction'). Section 105(2) of the Act excludes several activities for which the Green Book would seem most appropriate (assembly, installation or demolition of plant, machinery or supporting steelwork for chemical engineering process plants, water treatment works or power generation, etc. for example).

Summary of IChemE Green Book

Thus in summary, the IChemE Green Book provides a flexible mechanism for contracting in uncertain conditions. As a cost-reimbursable contract the risk of all costs associated with the construction activities lies with the Purchaser, and he is well-advised to consider employing one of the various cost-targeting methods available in order to limit his liability. In itself the IChemE Green Book does not allow for such a provision and thus could be seen as somewhat 'open-ended'. Collaborative relations between the Purchaser and Contractor are advised (especially given the Purchaser's exposure to costs), but there is no guidance how such relations should be established or maintained. This report advocates careful consideration should be given before employing this form of contract, in order to ensure the business objectives of the plant works can be successfully achieved.

Case Study 9: Thames Water Utilities Limited

Since the late 1980s, Thames Water has adopted the IChemE Green Book as one of its main forms of contract. Originally it formed the basis of Thames Water's 'extended arm' agreement with Taylor Woodrow Management Contracting Ltd, but it has also been used for design and build works for treatment and process works as well as other non-process works.

Over time, the contract has repeatedly shown to provide value for money to the firm. It adopts a collaborative approach to risk sharing which sits comfortably with what Thames Water is able to manage. Thames has amended the standard form by adopting target cost formulae to ensure the works are kept to budget; this has also had the effect of incentivising contractors to meet Thames water's targets.

The formulae are shown overleaf. Performance feedback has demonstrated that 60% of their contracts were within 5% of the agreed cost targets. When one considers that the majority of Thames Water's construction contains high levels of ground works in and around currently operational sites, this is an admirable benchmark for the rest of industry. There is little question that these figures can be improved on. Thames Water recognises that that the use of the IChemE is not for the risk averse client but, where flexibility and risk sharing is a pre-requisite of successful procurement, the Green Book provides a good overall framework.

> ### The Thames Water Target Cost Formulae:
>
> The contract price, $CP = ADf + AMf + PAc + BT$, where:
>
> ADf = the Actual Design fee which was tendered;
>
> AMf = the Actual Management fee (adjusted by a percentage of the actual costs);
>
> PAc = the Proportion of the Actual costs to be paid by the client, based on a proportion of cost sharing between the client and contractor;
>
> BT = a Bonus which only applies if timely Takeover occurs.
>
> Source: Walker, et al.[17]

Summary

This chapter has completed the review of common contracting methods and their standard forms of contracts in current use in the industry. It has examined the merits of the management approaches and performance contracts from the perspective of the inherent power structure within each of the contracting mechanisms.

Any prospective client of construction works would be well-advised to consider carefully the provisions of each of these methods as, within them, there are traps for the unwary. This chapter has shown that these contracts, like the others reviewed in Chapters 5, 6, 7 and 8, each contain a structure of power which once agreed, locks the parties in to their requirements.

Many of the industry's standard forms of contract appear to be structured against the client, in favour of the professional adviser. Perhaps this is unsurprising, as most of the contracts have been drafted by professional institutions wishing to protect and further their own members' interests.

These power imbalances are demonstrated in each of the key areas of control and works management within the contracts. Decisions to vary the contract, to expedite progress or to control the terms of payment are all made for the client without any recourse back to him. The client is locked-in with very little control or leverage on the transaction.

The expectation is that *successful* works contracts can only be delivered if the pre-contractual specification has been adequately and competently prescribed. Failure to accomplish this, or to maximise the commercial advantages in the pre-contractual negotiations, will only lead to project failure and compromises on business objectives in the post-contractual period.

The following section of this book considers public sector contracting and two of the standard forms of contract in use by Government and local authorities.

In the final section of the book the contracts examined in this section and the next are considered in the context of an over-determining business relationship and the structures of power which effect the commercial transaction and the way in which the business relationship can be implemented.

Chapter Notes

1. Bennett J., Pothecary & Robinson (1996)
2. National Audit Office (1996) *Progress in Completing the New British Library*, HMSO, London.
3. Dearle and Henderson (1988) *Management Contracting: A practice manual*, London, E&FN Spon, p. 2.
4. Refer to: Hayes R.W, Perry J.G. & P.A. Thompson (1983) *Management Contracting*, CIRIA Report No. 100, Construction Industry Research and Information Association, London.
5. Ward S. C, Curtis B. & C.B. Chapman (1991) 'Advantages of Management Contracting - Critical Analysis', *Journal of Construction, Engineering and Management*, Vol 117, No. 2 (June), *American Society of Civil Engineers*, pp. 195 - 211.
6. The Engineering and Construction Contract (1995) *Guidance Notes*, London, Thomas Telford, p. 15.
7. Refer to: Ward, *et al., Op. cit.*
8. Naoum S.G. & Langford D.A. (1987) 'Management Contracting – The client's view' *Journal of Construction, Engineering Management*, Vol. 111, No. 3, pp. 369 - 384.
9. Sidwell A.C. (1983) 'An evaluation of management contracting' *Construction Management and Economics*, Vol. 1, pp. 47 - 55.
10. Murdoch J. & W. Hughes (1996) *Construction Contracts: Law and Management*, 2nd Edition, E&FN Spon, London, p. 67.

11. Wright D. (1994) *An Engineer's Guide to the Model Forms of Conditions of Contract for Process Plant*, Institution of Chemical Engineers, Rugby.
12. Payne A. (1995) 'Comparing ICE6, MF/1 and the IChemE Green Book' *Proceedings of the Institution of Civil Engineers*, Vol. 108, Issue 4, pp. 189-191.
13. Institution of Electrical Engineers and Institution of Mechanical Engineers (1995) *Model form of general conditions of contract MF/1*, 3rd Revision, London, IEE and IMechE.
14. Wright, *op. cit.*
15. Nash P., R. McGill & J. O'Callaghan (1994) 'Procurement strategy and Contract management philosophy' *Proceedings of the Institution of Civil Engineers*, Vol. 102, Special Issue No. 2, pp. 76-82.
16. Perry J.G. & P.A. Thompson (1982) *Target and Cost-Reimbursable Construction Contracts"*, CIRIA Report R85, London, Construction Industry Research and Information Association.
17. Walker S.C.A. Remington R. & G.H. Bateman (1995) 'The Thames Water use of the IChemE Green Book' presented at the ICE Conference: *Major Projects: Can We Afford Them?* 5–6 April, ICE, London.

Part C

The Public Sector

Chapter 10

Public Sector Contracting

Introduction:

Why should this chapter and the next on Government Contracts be hived off from the rest of this book in a separate section of their own? The answer lies in the distinctive environment within which the public sector operates. In this chapter this environment is explored and shown to be considerably more complex to manage than that of the private sector. It is not just the presence of EU legislation for public works that effects construction operations, it is the contested nature of what is valued in the public sector and the need to meet probity standards for the use of public money that make the job of public contracting more problematic than that facing the private sector. Yet despite these seeming barriers to effective procurement management, public sector bodies have a major opportunity to gain effective leverage over the supply-base and deliver significant efficiencies and benefits with public expenditure. The key question facing them is whether they will grasp the nettle and implement a strategic approach to procurement, or whether they will prefer to keep to their current operationally tactical methods.

This chapter begins with an assessment of the problems in public sector management and proceeds to consider the effect of the existing regulatory framework in the UK. It concludes that there exists a confusion in terms of strategic clarity and, consequently, strategic direction for public sector administrators.

The chapter that follows this one proceeds to concentrate on two of the specific forms of contract used by civil servants to procure construction works. These are compared with the principles which Sir Michael Latham laid down for a modern contract and, as a result, concludes that there is a fundamental misunderstanding of business philosophy in terms of both the end-goals and the means by which they are achieved.

Problems in Public Procurement

To understand the issues associated with managing and regulating public expenditure, there are several key questions which need to be addressed:

- What are the end-goals of public spending and what are its 'values'?
- What are the most appropriate means of delivery and who should decide them?
- How should performance be measured?
- Who should play Solomon over allocative scarcity?
- What is the most appropriate regulatory structure to achieve valued outcomes?

These are incisive questions, for which there are no uncontested answers[1]. Each question is briefly considered under the following sub-headings and, while referring to public expenditure as a whole, has direct ramifications on the way construction is procured and managed.

The contested goals of public spending

It has been established in preceding chapters that the *raison d'être* of the private sector firm is the creation of a profitable margin which it can appropriate and accumulate for itself. There is, however, no clear overall intent for the public sector. Figure 10.1

indicates some of the possible end-goals ('valued outcomes') which public expenditure might be attempting to achieve. Doubtless there are others which could be added to this list, however what is clear is that there is no single defining purpose for which the public purse exists. Furthermore, several of these values are conflicting and, in order to satisfy all, there exists a tension at the heart of public sector management.

Much consideration has been given to the term 'value for money' as an end-state in itself, but this cannot be achieved unless the public sector values have been clearly defined. Since, as Figure 10.1 demonstrates, these values are manifold and contested, 'value for money' cannot, on its own, be a satisfactory goal to pursue. Rather, as contended in the following sub-section, 'value for money' is a means by which some of the goals in Figure 10.1 can be achieved. The problems associated with generic statements concerning value are predominantly subjective: i.e. value to whom and in relation to what?

SOCIAL GOALS:	• Protection of property and law & order;
	• National defence of territory;
	• Equality of opportunity;
	• Minimum levels of social infrastructure;
	• Redistribution of wealth;
	• Health gain and maintenance;
	• Environmental quality;
	• Minimum levels of social income;
	• Educations of labour force;
	• Protection against old-age and infirmity.
ECONOMIC GOALS:	• Higher levels of economic growth;
	• Protection of infant industries against competition;
	• Local / regional economic development;
	• Development of specific industries / technologies;
	• Investment in industrial / technological transfer;
	• Underwriting and subsiding national costs of production and exchange.

Figure 10.1 The Contested Goals of Public Spending.

The contested means of delivery

Given that there are so many disparate goals and objectives to fulfil, it follows that the means of achieving these goals will also be confused. Clearly the means of delivery will be contingent on the nature of the end-goal but, since this is confused, the ways in which public procurement should be managed effectively are also contested. As already mentioned, achieving these goals through 'value for money' (i.e. the correct valued outcome at the least expense) is fine as far as it goes, provided the correct measure of value is agreed and clearly identifiable. Where there is confusion or conflict, the means of delivery (and the very notion of 'value for money') cannot be determined. In this context, 'value for money' becomes relatively meaningless.

Furthermore, cutting across the public sector values identified in Figure 10.1, there will be the individual agendas of the ruling political party. This may be to minimise the role of central government by devolving authority, spending and public accountability, or it may be to consolidate power and expenditure in order to maximise the role of government. This agenda, overlying the contested ends and means of public spending, ensures that the framework within which public procurement is conducted is complex and, on many occasions, confused.

How is performance measured?

There is an adage which suggests that: 'what you measure is what you get'. The problem for public sector management is exacerbated by the range of views and advocates of the varying measures of success. Not only are the end-goals and the means to achieve them highly contested, since there is no agreement, the measures of good performance remain contested too (i.e. what is 'satisfactory' for one, may be contested by another).

There is also a question concerning from whose perspective the performance is measured (politicians, civil servants, the media, the electorate, unions, pressure groups, etc.). If, as is usually the case with publicly-transparent affairs, there are several factions, each with their own set of values and own measures of success, it is certain that public sector performance will continue to be contested.

Who should play Solomon?

Related to the above problem is that of deciding who should manage and regulate the demand for valued outcomes when conditions of scarcity prevail. By this is meant that the demand for public expenditure continues to outstrip supply. The level of demand for valued deliverables from the public purse is potentially infinite (in terms of both articulated and unarticulated needs and wants). Much is found wanting when these are balanced against the available level of resources that exist. In the contemporary political climate, where the electorate prefers to reject increases in taxation, resources are relatively scarce for these services. Nobody wants to pay for the high level of services that are demanded.

The key questions concerning public expenditure is who should be deciding where the scarce resources are to be allocated? Despite the political rhetoric, it is clear that there are those who will benefit and those who will lose. This is just as much the case in construction as it is across the whole of the national economy.

What regulatory structure is appropriate?

It follows, therefore, that there is distinct confusion within public sector management, irrespective of which political party governs. The confusion between means and ends and the contested nature of public values ensures that, for any regulatory framework, it is unclear whether it can achieve its intended effect, since this is contested too. Two propositions can be drawn from this:

1. There is normally a complete confusion in public regulatory structures over whether the primary goal is either to ensure the probity of public spending, to bring about desired economic and/or social goals, or to achieve 'value for money'.
2. In practice, most regulatory regimes, because they are written by politicians and civil servants, focus primarily on probity and social and economic goals. This leads to confusion and makes

it extremely difficult to focus on value for money and expenditure efficiency.

These propositions can be verified in the light of the EU Public Procurement rules which have a direct impact on the way in which construction is procured by Government. However, as discussed in the following chapter, the UK Governement is not a monolithic organisation and, despite possessing the capacity to consolidate expenditure for greater leverage and efficiency in the market place, it has, in the recent past, preferred to offer suppliers fragmented expenditure tendered on a once-off price-only basis, in an *ad hoc* and piecemeal manner.

Moreover, the current regulatory structures are applied holistically across all industry sectors, irrespective of the contextural circumstances. There is very little long-term thinking; expenditure is driven by annual budgets and changes in administration at the national, regional and local levels.

It is contended in the remainder of this chapter that this environment militates against effective procurement management. Moreover, it can be argued that the presence of the EU Procurement directives (and, equally important, the manner in which the public sector has interpreted them) provides a structural barrier to more effective procurement management. It is little wonder that the Levene Report observed that the UK Government is not a best practice client of construction works[2].

The EU Public Procurement Rules

There are four main directives for public procurement in the EU (supplies, works, services and utilities) which came into effect between 1988 and 1992. Overall, the legislation had two major goals:

1. to end national preferences in the award of public sector and utility contracts and
2. to assist in the removal of artificial barriers to trade by encouraging greater cross-border trade in the award of public sector and utility contracts to second and third country suppliers.

If this could be achieved, it was contended, public expenditure across the whole EU would be reduced and the industrial supply-base of Europe would become more competitive. However, on the evidence to date this is not occurring[3].

The EU legislation is based on classical neo-liberal economic thinking. It was the ideology behind the Cechinni report and, more specifically, the Atkins report which suggested that the legislation should create more transparency and liberalise the award of contracts to non-national suppliers in order to 'kick-start' a chain reaction of higher levels of economic growth throughout industry. This reaction, based on the concept of open competition in the 'perfect' market, has three principal mechanisms beholden to it:

1. The *static price effect* suggests that, in the short-term, suppliers hold to their prices and the lowest-price bidder in an open market (across national boundaries) gets the contract awarded.
2. The *competition effect* suggests that, in the medium term, competition will begin to increase as other suppliers begin to push prices down and become 'more competitive' as a response to losing contracts through static pricing, therefore offering greater efficiency for the public purse.
3. The *restructuring effect* which suggested that, in the longer term, there would be a fundamental restructuring in key industrial sectors dominated by public purchasing. This would result in the inefficient suppliers exiting the market and leaving the larger players (with greater economies of scale and efficiencies) to provide yet lower prices.

The Atkins report suggested that savings of between 8 and 19 billion ECUs (i.e. £5 – 12 billion) at 1984 prices, while at the same time 400,000 additional jobs could be hoped for after the restructuring insolvencies were lost. Recent research[3] demonstrates this has not occurred and seems unlikely to do so.

The test for the success of this approach would clearly have been a significant increase in the number of cross-border contract awards to non-national suppliers (one of the original objectives of the legislation and a clear indication that the static price effect was operating across national boundaries). However recent research indicates that European public and utility markets were more open than the EU's initial thinking presupposed and that little cross-border trade has occurred since 1992, as most large MNCs had already located subsidiary offices in other European states. Furthermore the rules are framed on the basis of a fundamental misunderstanding of how supply markets operate and what the cost drivers and power structures are within them. Despite the evidence of failure, in their review of the public procurement rules in 1997, the EU's Green Paper has continued to reinforce its neo-liberal thinking.

The intellectual myopia behind the rules is, on reflection, clear for all to see. They are forcing public and utility bodies into the straight-jacket of formal contracting procedures with a large number of suppliers. This is predicated on assumptions that increased competition will encourage lower prices and greater efficiency.

This is the complete antithesis of most private sector thinking about what is 'best practice' in procurement management. On the contrary, privately-owned companies are recognising that an increased number of suppliers can be wasteful and inefficient. Supply-base rationalisation has demonstrated that managed collaboration with a limited number of suppliers in competition saves transaction costs and provides greater consolidation of purchasing spend (i.e. incentive) among suppliers to induce preferred supply offerings.

The results in the private sector have indicated cost savings, greater efficiency, improved delivery, enhanced quality, less waste and overall better supply co-ordination. By taking the opposite approach, in the face of the opportunity to consolidate considerable purchasing spend (whether in construction or other industrial sectors), the EU has elected to ignore these benefits in preference for the pursuit of simplistic ideological goals which appear to be failing to work. The effect of this regulatory approach may, in fact, lead to increases in transaction costs throughout the public sector, at the expense of the EU tax-paying public.

Public Sector Construction Procurement

An example of the dysfunctionality of simplistic neo-liberal thinking is provided by UK approaches to public sector construction management. The UK Government was once the largest construction client in the UK. However, successive policies of devolvement, privatisation and market testing have fragmented the purchasing power of the public purse. Construction procurement used to be co-ordinated through the Property Services Agency but, as reported in Chapter 11, authority for expenditure has been dispersed among 40 Government Departments, over 100 'Next Step' Agencies and the 400 or so NHS Trusts. The result has been a considerable loss of *purchasing power* in the market.

In this context, *purchasing power* means the relative ability to exert leverage and/or incentives over the supply-base in order to realise one's own objectives and goals. By fragmenting its expenditure in construction, the UK Government has lost relative ability ('clout') to shape and direct the service delivery it receives from the industry. Thus, the public purse is subject to the opportunistic supply offerings of the privately-owned construction supply-market without any real or effective ability to influence or change performance. In supply chain management terms, it has become a [reactive] price-receiver, rather than a [proactive] price-fixer.

Construction works are, as a result, likely to be contracted in piecemeal fashion on an *ad hoc* aggressive tendering basis. Furthermore, the government has erected other barriers to effective procurement. These include strict budgetary controls and procedures, annuality of expenditure and blanket budget cuts (e.g. 5% year-on-year reductions). The effect has been to ensure public construction procurement is focused on compliance with the procedures, rather than the performance-outcomes they deliver. In this culture, it is better to spend the budget than to effect annual cost savings below it. It lends itself to a 'quick-spend' mentality in the last quarter of the budget to ensure all allocated resources are used up. This type of practice is wasteful and does not encourage cost-reduction or enhanced supplier performance.

Evidence of the poor construction procurement practice in the public sector abounds. In 1997, the Audit Commission published the results of its review of 60 local authorities' capital expenditure[4]:

- 25% of the 700 projects reviewed were over-spent by 5% or more;
- Of the projects reviewed, more than two-thirds ran late and half of them were 15% or more behind the originally-scheduled duration;
- 95% of the projects reviewed included contingency sums of between 6% and 14% on average in addition to the contracted sum;
- professional fees ranged from 2% to 43% of the prime cost sums;
- cost estimates for 50% of the projects reviewed are not updated between the initial budget estimate and the actual scheme design;
- 25% of the local authorities reviewed only plan for one fiscal year at a time;
- 50% of the local authorities reviewed had no readily accessible details on the condition of their property assets and, of those who had a database, less than 40% actually used it to help plan their capital expenditure.

The evidence speaks for itself and, typically, represents the generally low level of competence in public sector construction procurement.

This is not to suggest that, by consolidating construction expenditure under the jurisdiction of a single public entity, the degree of procurement management and *contracting competence* would necessarily be improved. It is simply to suggest that, in so doing, the public sector would remove the structural barriers to effective procurement management and the quantum of possible efficiency gains and performance improvements would be greater. This would also have the effect of consolidating procedures and practices, thus making the implementation of better practice more readily possible. Indeed, better practice procurement management might encourage the public sector to consider insourcing, rather than contracting out, significant areas of public works where the public sector is itself the primary client in the country. This would be a form of supply chain management by vertical integration.

Summary

This chapter has examined the background to public sector contracting. It has demonstrated that public procurement management is considerably more complex than that within the private sector. Notwithstanding this, this chapter has exposed the poor levels of competence and ineffective management that currently exist. The three main reasons for this have shown to be:

1. the contested nature of public values which has been demonstrated to add confusion as to which means of delivery and performance measures are effective;
2. the intellectual myopia at the heart of the EU procurement legislation which has had little or no effect on the construction market;
3. the fragmentation of spending authority by recent UK governments which has removed purchasing power from the public administrators thus making the public purse subject to the supply-offerings of the market without any steer on performance improvement.

To effect good practice construction procurement, a fundamental and holistic review of the roles, responsibilities, accountabilities and authorities of all public spenders is required, in order to examine ways in which the loss of control and purchasing power can be redressed and the general level of contracting competence can be raised throughout the public domain.

The following chapter takes a detailed look at two of the standard forms of contract used by Government. Like the previous chapters, it examines the structure of power and allocation of risk within these Government contracts, as well as considering the efficacy of their underlying business philosophies.

Chapter Notes

1. Refer to: Cox A. (1993) *Public Procurement in the EC Volume 1: The Single Market Rules and the Enforcement Regime After 1992*, Earlsgate Press, Boston, UK; Cox A. (1995) 'Strategic Procurement Management in the Public and Private Sectors: The Relative Benefits of Competitive and Collaborative Approaches' in Lamming R. & A. Cox (eds.) *Strategic Procurement Management in the 1990s: Concepts and Cases*, Earlsgate Press, Boston, UK; and Furlong P. & A. Cox (1995) *The European Union at the Cross-Roads*, Earlsgate Press, Boston, UK.
2. Levene P. (1995) *Construction Procurement by Government: An Efficiency Unit Scrutiny*, HMSO, London.
3. Cox A. & P. Furlong (1995) 'Utilities Contracting and the EU Procurement Rules' *Utilities Policy*, Vol. 5, No. 3/4, pp. 199 - 206; Cox A. & P. Furlong (1995) 'European Procurement Rules and National Preference: Explainingthe Local Sourcing of Public Works Contracts in the EU in 1993' *Journal of Construction Procurement*, Vol. 1 No. 2 (November), pp. 87 - 99; Cox A. & P. Furlong (1996) 'The Jury is still out for Utilties Procurement' *Public Procurement Law Review*, No. 5, pp. 57 - 66.
4. Audit Commission (1997) *Rome Wasn't Built in a Day*, HMSO, London.

Chapter 11

Government Contracts

Introduction

This chapter follows on from Chapter 10 by considering the principles and power structures inherent in two forms of contract which have been published for use by government departments and other quasi-non-governmental organisations. These forms of contract are:

- the GC/Works/1 Edition 3; and
- DEFCON 2000.

These contracts are not used by all UK public sector bodies; indeed many Government Agencies use the industry standard forms described in the previous Part of this book. Nevertheless the following contracts, referred to as 'Government Contracts' here as they were specifically drafted for Government, remain popular and commonly used both within and outside of the public domain.

Background

The UK Government was once a 'monolithic' client of construction works centrally co-ordinated through the Property

Services Agency (PSA), which was responsible for the development of the GC/Works/1 contracts which have been widely used throughout government.

The need for a government contract had first been recognised during the Second World War in order to enable national construction projects to be completed using relatively scarce resources and yet without delay. Government departments had been using the 1939 RIBA Form and/or their own forms of contract, but none suited the specific needs of the time. As a result, the CCC/Works/1 contract was unilaterally developed for government in 1943 and subsequently ran into nine editions. However as other standard forms of contract were developed after the war (the *ICE Conditions of Contract 1st Edition* was published in 1945 and, in 1955, the Joint Contracts Tribunal was widened to include representations from local authority associations), the scope of need for a government standard form of contract narrowed. The GC/Works/1 was published in 1959 and, as it gained popularity, so the CCC/Works/1 was phased out. Subsequently the GC/Works/1 was championed by the PSA which resulted in a second edition in September 1977 and the third edition in December 1989, toning down the punitive contract terms that had been required during the war-effort.

The GC/Works/1 forms of contract have been widely accepted and used throughout government with much satisfaction. Indeed one commentator has observed: *"...it is undoubtedly the best standard form of building contract available and shows all the advantages of unilateral provenance. Since it is drafted on behalf of the Department of the Environment it is more clear cut than the negotiated forms such as the JCT Standard Form JCT 80 and the Intermediate Form IFC 84 which are often an unhappy compromise. Edition 3 approaches the allocation of risk in a sensible and businesslike way"*.[1]

However following the demise of the PSA, initially in 1988 and more generally in 1990, government departments have been made responsible for their own capital expenditure budgets. This has resulted in approximately 40 central government departments and over 100 'Next Step' Agencies becoming 'new' clients of construction. At a similar time the National Health Service (NHS) was being decentralised creating more than 400 NHS Trust clients. This fragmentation of expenditure has had the effect of diversifying policy and practices throughout the UK Government. Furthermore, it has recently been considered that: *"...Government is not yet getting all-round value for money, nor is it a best practice client. Even its best projects and systems have substantial room for improvement."*[2].

The Latham Review (commissioned by the Department of the Environment) was aimed at trying to resolve some of these issues, in particular the contractual and procurement arrangements in the UK construction industry[3]. Among his recommendations, Sir Michael Latham promoted the broad use of the NEC. However there has been a sense of unease in many sectors of the industry that this form of contract is not the optimum for all circumstances, as discussed in Chapter 8.

Despite Latham's preference for government to widely adopt the NEC, the Levene Report subsequently observed that many Departments *"...are strongly opposed to it"* [paragraph 249, p. 75]. Furthermore, it reported that some government departments, the Ministry of Defence (MOD) being one, had conducted independent reviews of the standard forms of contract and had maintained their preferences for the GC/Works/1.

Among this throng of NEC-sceptics stands the MOD's Defence Estate Organisation (DEO), the UK's largest client of construction, which, as reported in the trade press, has rejected the use of the NEC[4]. Instead of accepting Latham's principles for a 'modern contract', the DEO (in consultation with industry) continued to pursue its programme of modifying the GC/Works/1 contracts to provide it with its own forms of contract, called DEFCON 2000.

In short, DEFCON 2000 comprises a family of contract forms, based on a modified version of GC/Works/1 (3rd Edition), which were prepared in the course of 1995 and 1996 in consultation with the industry. The contracts have been initiated by the DEO which the MOD established in April 1995 to provide it with an in-house resource of professional works advice, a contract-awarding body and an interface between itself and the construction industry.

This chapter begins with an examination of the DEO's new contracts and their prospective utility in an industry that is already awash with many different industry standard forms of contract. The contents are based on copies of the agreed final draft of DEFCON 2000 and privileged discussions with senior staff at the DEO, as part of the authors' research agenda[5]. The chapter then proceeds to consider the GC/Works/1 contract as the preferred government contract.

In the course of reviewing these contracts, some important yet fundamental questions have had to be asked:

- are government contracts 'fit-for-purpose' for construction?
- is the industry willing to accept yet more standard forms of contract?
- will further contracts incur additional costs of transaction? (And if so, are the extra costs worth it?)
- what was wrong with Latham's proposals and the NEC?

DEFCON 2000 [6]

The 'DEFCON' title is not as formidable as it first appears; the term relates to the MOD Procurement glossary of contract clauses, known as the Defence Conditions, of which, at present, there are approximately 200 in number. DEFCON 2000 is the standard set of conditions for the DEO, which will take it into the 21st Century. Its intended use is for all capital projects over £240,000 (exclusive of VAT and fees) both in the mainland UK and on international military bases (such as Ascension Island). At present, there are four main contracts:

- a sequential contract;
- a design-build contract;
- a works subcontract; and
- a contract for the appointment of a project manager.

This chapter concentrates on DEFCON 2000's design-build version: the 'Works Contract with Contractor's Design: Conditions of Contract' (Agreed Final Draft, October 1996).

In short, it is a lump sum, single point responsibility, design and construct contract based on a milestone payment system and an express fitness-for-purpose requirement on the design. As can be expected from the MOD and indeed any contract based on GC/Works/1, DEFCON 2000 carries an air of military crispness in its drafting. The industry's compliment of 'tough but fair' (formerly given to GC/Works/1) will no doubt be levied on this contract too as the risks and responsibilities have been very clearly defined (indeed, it may even be considered to be 'tougher but fairer'!). Although rigorous, DEFCON 2000 clearly states the roles, responsibilities and risks within its express terms. The aim, according to those responsible for its innovation, is to avoid unnecessary 'surprise' disputes arising from a lack of clarity regarding who is responsible for what (or in other words, to let contractors know what to expect up-front). Examples of this approach can be seen in the following:

- **Design Liability** – The contract specifies an express fitness-for-purpose requirement on works designed by the contractor: *"Subject to the provisions of paragraph (5) and (6) of this Condition, the Contractor warrants that the Works shall be fit for the purpose or purposes described in the Design Brief."* (Condition 19(4)).

  These exceptions, in paragraphs 5 and 6, refer to instances where the fitness for purpose requirement may be reduced: *"Where the state of scientific and/or technical knowledge in relation to any part of the Design Brief at the relevant time was such that it was not practicable or feasible for the Contractor to warrant the fitness for purpose of the whole, or any part of the Works, the Contractor's design liability in relation to that part of the Design Brief shall be to exercise the skill care and diligence to be required or expected of a properly qualified and competent designer or contractor experienced in carrying out work of a similar nature, scope and complexity......However, the Contractor shall have a continuing duty to keep any such design under review during the Contract up to and including the date of the last Maintenance Period to expire and to notify the Project Manager in the event that the Contractor gains any knowledge that suggests the Works or any part thereof may not be fit for their intended purpose."* (Condition 19(5)).

This assumes that the Design Brief has clearly expressed the intended purposes of the finished works. Furthermore it is not made clear who is to decide that a fitness for purpose requirement is 'not practicable or feasible'; will the DEO's Design Briefs expressly state this up-front, is the contractor expected to make a claim, or will this only be decided retrospectively before tribunal?

The other exception refers to *"any part of the Design Brief containing a Prescribed Design Solution"* (Condition 19(6)), which is essentially works designed by the client and prescribed in the tender documentation. The Contractor is required to accept the prescribed works and be responsible for their development detailing and co-ordination; but his liability is reviewed and 'reduced accordingly'. Similarly, the liability for the design of variation works is apportioned between that which is fit for purpose and that which carries a reasonable skill and care requirement.

These conditions, although somewhat laboriously spelled out (in 16 sub-clauses), are no more stringent than that required under common law. The fact that the liabilities are spelled out reinforces the DEO's claim that fairness lies in a contract's clarity to define risks and responsibilities. In this regard DEFCON 2000 compares favourably with its contemporary standard forms of contract (see, for example, clause 2.5.1 of the JCT'81).

- **Ground risks** – DEFCON 2000 draws the distinction between *artificial obstructions* and *ground conditions*. Because ground conditions (both geological and geotechnical) can be reasonably foreseen and predictable through various methods of site investigation, DEFCON 2000 apportions this risk to the contractor (Condition 7(4)(d)). The Authority (client/employer) warrants the accuracy of any factual information passed to the contractor; so that where the risk is not reasonably foreseeable (e.g. in the case of an unrepresentative borehole, *inter alia*) the risk is excepted: *"Where the Authority has made available to the Contractor before submission of the Contractor's Tender factual information on the ground and sub-soil conditions* *obtained by, or on behalf of, the Authority in relation to the Site, including the location and extent of known Artificial Obstructions, the Authority shall in the Design Brief warrant the accuracy of that part of such factual information which has been obtained by or on behalf of the Authority from investigations undertaken for the purposes of the Works"* (Condition 7(1)).

If the contractor is responsible for conducting the geotechnical investigation himself, then he is expected to have covered this either through his own insurance or a risk premium in his tender. At first sight this may appear onerous, but the view of the DEO is that this is a statistically predictable event and therefore can be managed as such. The contractor is better placed to accept the risk as this is (or at least should be) one of his core competencies.

However in the case of artificial obstructions, the DEO is prepared to share the risk with the contractor. Most works are conducted on MOD operational sites and thus, in the cases where obstructions are known and the contractor is forewarned in the invitation to tender, the contractor is given responsibility for the risk (Condition 7(4)(e)). However when the obstruction is unknown or unforeseen, or the information provided about it is inaccurate, the Authority accepts the risk (Condition 24(1)(a)(v)).

- **Adverse Weather** – DEFCON 2000's policy on adverse weather is not fully known as Appendix 4 'Weather' (compiled in association with the Meteorological Office) was not included in the Revision 1 draft consultation copy of DEFCON 2000 and has been omitted from the Agreed Final Draft (on which this chapter is based). However, from discussions with staff from the DEO, it is believed that the intention is to statistically monitor weather conditions (rain, snow, frost, etc.) on each site as all MOD sites have Meteorological Office weather stations. 'Adverse' weather (for which the Authority will accept the risk) is then measured as +/− 3 standard deviations against the mean. Again it is expected that the contractor can statistically manage the risks of inclement weather within this range, with the client covering only the most extreme of conditions.

These allocations of risk may be seen as controversial and, no doubt, there will be some (particularly on the supply-side) who will raise dissent against the hardship of DEFCON 2000. Indeed the DEO are not expecting other clients to rush to take up its use. Nevertheless the risks are clearly identified and apportioned (Condition 24 even lists all the Authority's risks) and, as such, their 'fairness' is plain for all to see: both pre-tender and post-award. It is unfortunate that the same cannot be said of all standard forms of contract. Indeed, it is the vaguities and lack of precision of some of the agreed standard forms, which can then be exploited opportunistically, that some consider to constitute an 'unfair' contract.

Notwithstanding this, DEFCON 2000 has elected to retain one of the most criticised clauses of government contracts: the express rights of set-off against any other government contract. Indeed this provision is almost an exact replication of Condition 51 of GC/Works/1: *"Without any prejudice to any other rights of the Authority whenever under the Contract any sum of money shall be recoverable from, or payable by, the Contractor such sum may be deducted from the amount of any sum or sums then due or which at any time thereafter may become due to the Contractor under, or in respect of, the Contract or any other contract with the Authority or with any department or office of Her Majesty's Government."* (Condition 47).

Like its GC/Works/1 predecessor, DEFCON 2000 is a 'project-orientated' contract suitable for procuring capital works from the public purse. Some of its project management features include:

- **Project Manager** – as with most standard forms of contract, DEFCON 2000 is based on *trilateral governance*. Like both the NEC and GC/Works/1, this is not biased to any one professional institution; DEFCON 2000 makes no reference to a Quantity Surveyor. Thus the client's main appointment is the Project Manager who is required to possess the appropriate skills and competencies required to serve the client's needs in the contract. This may only seem to be a small measure, but its effect could be considerable for the client who wishes to remove unnecessary tiers of 'professional' cost from the management of a works contract.

- **Programme** (Condition 34) and **Progress Meetings** (Condition 36) have been prescribed to assist with the expeditious completion of all the works and any issues arising from them. These clauses are very similar to the provisions of Condition 33 (Programme) and Condition 35 (Progress Meetings) in GC/Works/1, mentioned later.

- **Authority's Risks** – as already mentioned, these are listed in Condition 24. They go further than the 'Excepted Risks' commonly found in other standard forms of contract, and list all risks accepted by the client.

- **Project Management Procedures** – to assist with some of the procedural elements of contract administration, DEFCON 2000 has appendices for the administration of: Change Control (Appendix 1) and Payment (Appendix 2). Although these spell out the principles for the management of the project regarding these specific issues, they are long and not readily comprehensible without careful scrutiny. In this regard DEFCON 2000 would perhaps benefit from a lesson from the NEC and its use of flow-charts to illustrate actions and procedures (particularly the Change Control appendix).

- **Guidance Notes** – it has been recognised that this contract is unlikely to be covered by many commentators in forthcoming literature and so it is believed that the DEO will be publishing Guidance Notes to accompany the contracts.

- **Social Clauses** – like the GC/Works/1, DEFCON 2000 contains a number of clauses addressing particular social aspects of working on government schemes:

  - Prevention of Nuisance and Pollution (Condition 12);
  - Prevention of Damage to Public Roads (Condition 13);
  - Protection of Personal Data (Condition 14);
  - Prevention of Racial and Sexual Discrimination (Condition 16);
  - Prevention of Corruption (Condition 17); and

- Fair Employment (Northern Ireland) Act 1989 (Condition 62).

Finally, it is worth discussing the DEO's future intentions for DEFCON 2000. Following its trials on a few hand-picked projects in the course of 1997/98, it is expected that DEFCON 2000 will be broadly adopted by the DEO in place of GC/Works/1. However it is also believed that the DEO has a hidden agenda to take the contract further, in an attempt to re-shape the construction supply market.

One suggestion that has been proposed is to amalgamate the Project Manager's contract with the Design–Build Works contract to create a 'total package' *Design–Project Manage–Construct* contract. Thus the main contractor would become the only interface with the client: a true single point responsibility. This proposal has been called the 'prime contractorship' system of procurement which, when combined with operational and maintenance responsibilities, will create a totally integrated first tier in the construction supply chain, with the aim of confronting the industry's blight of fragmentation.

But will this work? The prime contractorship system's success would seem to rest with the ability of the supply market to re-engineer itself to the demands of the client-side. But this will only be effective if either a considerable market share of clients accept this system of procurement which, as already suggested, is unlikely, or the DEO gives most of its work to one or two prime contractors, which also seems unlikely given the arms-length adversarial intent behind DEFCON 2000. Furthermore, it offers little hope to the rest of the supply chain in which the greatest degree of fragmentation occurs. It is only a truly integrated and 'lean' supply chain which can offer the end-consumer greatest efficiencies and cost savings; which will only occur when regular construction clients re-evaluate their portfolio of construction expenditure, as suggested by Cox & Townsend[7].

It would appear that DEFCON 2000 simply offers the industry another alternative contract with which to conduct business at arms-length. The suggestion in this chapter is that DEFCON 2000 simply reinforces the adversarial attitudes and fragmentation which already exist in the industry.

GC/Works/1

The GC/Works/1 carries the full title of *"General Conditions of Contract for Building and Civil Engineering"* and has been developed by the Department of the Environment (DOE) for use with all government-sponsored construction works, whether civil engineering or building. The most recent version is Edition 3, published in 1989 by HMSO and revised in 1990. It is a lump sum sequential contract which, similar to the ICE 6th Edition, provides the client with an opportunity to specify some partial design input from the contractor. There is also a separate design and build version available.

The government's recent Efficiency Unit scrutiny (by Sir Peter Levene)[8] reported that the GC/Works/1 was regarded by the construction industry as *"tough but fair"*, and concluded that it *"...should on no account be dropped until there is something better to put in its place, with a proven track record"* [paragraph 251, p. 76]. This last comment is either a veiled reference to the NEC or DEFCON 2000.

One of the most striking features of the GC/Works/1 is its simplicity of language. The terminology is precise and does not leave decisions open to the subjectivity of third parties (unlike the JCT or ICE contracts). Furthermore roles and responsibilities are clearly defined. The client is referred to as the 'Authority' (perhaps the only concession to government traditions in the contract) and is represented through the contract by a 'Project Manager' (PM) acting in an agency role.

The PM is appointed by the Authority for the purpose of managing and superintending the works (Condition 1(1)) and therefore is likely to be either an architect (for architecturally-led building works) or a civil engineer (for civil engineering or infrastructure works). The Authority also appoints a Quantity Surveyor (QS) who is responsible for measuring and valuing the works, thus releasing the PM to concentrate on the contract management.

The PM's powers are wide, and include:

- appointment of assistants and the delegation of authorities to them (Condition 4(2));
- giving consent to sub-letting the contract (Condition 62(1));
- the correction of ambiguities, issue of additional drawings and revision of information (Condition 40);
- the agreement of the works programme (Condition 33(2));
- the removal of the contractor's personnel from site (Condition 6);
- instructions for exploratory investigations (Condition 40(2)(I));
- instructions for tests and/or investigations on materials and workmanship (Condition 31(4));
- removal and re-execution of unsatisfactory work (Condition 40(2)(d));
- suspension of the works (Condition 40(2)(g));
- the award of extensions of time (Condition 36);
- instructions to vary work requirements (Condition 40);
- the certification of payments and final account (Condition 50).

These are wide ranging powers which remove the Authority from the day to day operations of the contract, placing them in the hands of the PM. The Authority would be wise to ensure it carefully selects the most appropriate PM with the full complement of competencies required to undertake these tasks, and that in this undertaking it ensures the PM is appropriately liable for his actions when representing the Authority. This issue of liability is not covered in the GC/Works/1 documentation.

The GC/Works/1 contract has been described as 'project orientated' and contains a number of key features which promote good project management. The promotion of good management has been one of the key marketing drives of the NEC, as was discussed in Chapter 8. However, unlike the NEC, the GC/Works/1 does not prescribe the procedures that should be adopted in the course of the administration of the contract. The parties are left to manage the project using their own expertise enabling them to maintain their focus on performance rather than procedure. Some of these project management features include:

- **Programme** – Condition 33 requires the contractor to submit a detailed programme showing not just the sequence of work and proposed resources, but also temporary works details and method statements. (One omission however is guidance on corrective actions in the event that the contractor's programme is unacceptable to the PM).
- **Progress Meetings** – Condition 35 requires that monthly progress meetings should be held to assess the progress of the works and facilitate their satisfactory completion. Specific actions are required of both the PM and the contractor prior to the meeting to ensure there is sufficient flow of information to effect progress on all outstanding items.
- **Principled valuations of Instructions** – The contract stipulates governing principles by which the QS will value the PM's instructions. This includes an option for the contractor to provide an up-front quotation before commencing the works.
- **Cost savings** – Condition 52 allows the contractor to use his expertise to propose cost reducing methods to the PM, on the basis (and incentive) of sharing the savings equally between the Authority and contractor.

The reimbursement structure of the GC/Works/1 was totally revised when Edition 3 was published. It remains a lump sum contract, but there are two versions: that with quantities and that without. The advantage of the quantities version is that measurement (particularly of variations) is easier to determine and should therefore be employed for works where there is a greater level of uncertainty (unforeseen conditions, ground risk, etc). The works are re-measured (Condition 18) and if found to be discrepant with the Bill of Quantities, the Contract Sum is amended (Condition 3).

For both the 'with' and 'without' quantities forms of GC/Works/1 the interim and final payments are made on the basis of the Contract Sum. Interim payments are made monthly (as 'Advances on Account', see Condition 48) as a proportion of the Contract Sum based on the percentage of work done to date. This is calculated using the appropriate stage payment charts for the contract. Figure 11.1 illustrates the chart for works where the contract sum exceeds £5.5 million. The distinctive 'S'-shape may be representative of the general 'theoretical' trend of progress on site (i.e. slow at first, full progress mid-way through, and tailing-off towards the end), but it does not represent the contractor's off-site works or his own expenditure, which is typically heaviest up-front. Therefore, the contractor is either financing the initial works himself (between placing orders and receiving reimbursement from the client), or placing the burden on his suppliers and sub-contractors. Either way, this will be represented in additional costs to the client's final cost.

While a fixed reimbursement curve may be easy to administer and prevent 'front-end loading' on tenders, it could prove counter-productive in that it 'dis-incentivises' the contractor from progressing the works expeditiously: i.e. the contractor could wish to leave his most expensive/costly works to the end of the contract period thus creating a 'last-minute dash' to completion.

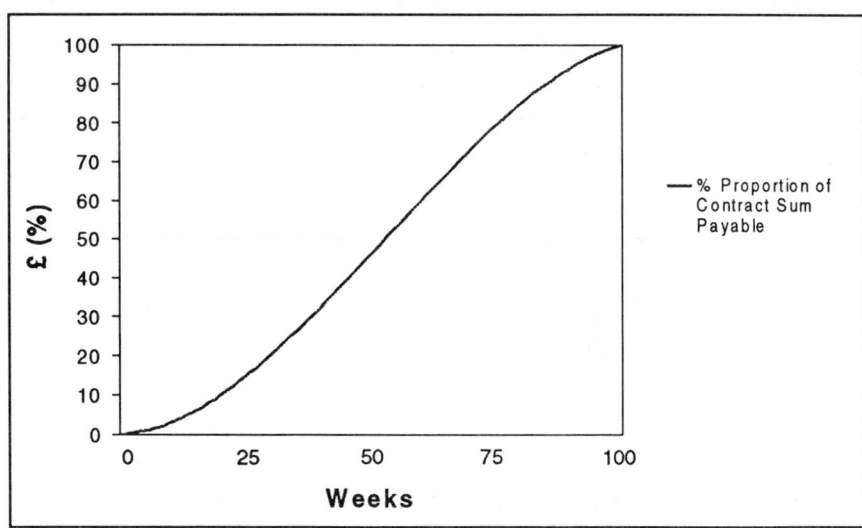

Figure 11.1: Stage Payment Chart for works > £5.5 million
Source: Potts[9]

Another measure for concern in the GC/Works/1 is the adjudication clause (Condition 59), which allows any matter in dispute (provided it has been in existence for more than 3 months) to be taken to the adjudicator. The contract states that the dispute (provided it has been in existence for more than 3 months) to be taken to the adjudicator. The contract states that the adjudicator should be a nominated officer of the Authority or person acting for the Authority and therefore can hardly be seen as impartial. The PM and QS have 14 days to submit any representations they may wish to make. From the receipt of the contractor's notice of adjudication, the adjudicator has 28 days to form a binding decision (for which he is not required to give any reasoning). In addition to adjudication, disputes may be referred to arbitration (Condition 60) where there is a maximum 6 month period for the hearing, followed by a 3 month period for the arbitrator to make his award.

It should be noted that this adjudication clause is in conflict with the Secretary of State's 'Scheme for Construction' (Part 1: Adjudication) of the Housing Grants Construction and Regeneration Act 1996.

Where the client is not satisfied with the clauses of GC/Works/1 he has provision to incorporate special conditions of his own. These should be listed in the abstract of particulars prepared in the invitation to tender and might include provisions for, for instance, the CDM Regulations, which have not been catered for at present.

To summarise, despite the queries expressed over the fairness of the fixed reimbursement curve and the partiality of the adjudicator, the GC/Works/1 remains a popular form of contract. It has been described as 'tough but fair' and is well-written in easily understandable English, clearly defining roles and responsibilities. It is not biased to favour any single professional institution, but is instead orientated to promote good project management in general.

For those looking for further information and reading, there is only a limited source of available literature. One recommended commentary is: Vincent Powell-Smith's *GC/Works/1 Edition 3 The Government Conditions of Contract for Building and Civil Engineering*, published by BSP Professional Books, Oxford in 1990.

A Critique of the Government Contracts

This chapter began by posing four fundamental questions of government contracts: are they fit for purpose, will the industry accept new contracts such as DEFCON 2000, will those new contracts incur additional cost and what was wrong with Latham's proposals for a Modern Contract? Although it is not possible to pre-empt the intentions of the government's contract drafters, some suggestions are offered in the following.

Fit-for-purpose contracts?

The DEO spent several years reviewing the contractual practices of the UK construction industry and developing DEFCON 2000. The natural extension of this assumes this contract will fully serve the DEO's purposes. When compared with other standard forms of contract, both DEFCON 2000 and the GC/Works/1 possess refreshing clarity. They are aimed distinctly at ensuring contractual compliance and good project management. However, despite the current vogues for trust and co-operation, both forms of contract are non-collaborative; in today's world of public sector accountability and probity, the government continues to ensure that it will hold its contractors at arms-length. This may suit the convenience of auditors, but will it really produce Value for Money procurement? As the UK's largest commissioner of construction works, surely there is considerably more benefit to be gained by taking a 'serial' or 'process' approach to procurement and working closely with an optimum number of preferred suppliers in competition?

Will DEFCON 2000 be accepted?

Just the briefest of glances through some of the trade press articles and letters will quickly remind the reader of the difficulties experienced by the NEC during its introduction and acceptance as an industry standard form of contract. With the plethora of ICE and JCT contracts already in existence, or in the process of being drafted, as well as other forms (such as FIDIC, IChemE and GC/Works/1) the need for yet more contracts is surely questionable. Nevertheless, it is probably inevitable that contractors will be willing to accept DEFCON 2000's terms; as already mentioned, the DEO is the UK's largest procurer of construction with an approximately 4% market share and, in present conditions, suppliers are grateful for all the work they can find. The downside, of course, is that when the industry experiences an upturn (such as the one offered by the turn of the Millenium) the 'boot' may be on the other foot and it will be the DEO who may rue its arms-length and adversarial approach.

Will more contracts incur additional cost?

There still exists a naive belief, in some quarters of the industry, that the best way of contracting in the industry is to write a bespoke contract to cover the precise requirements of the particulars of the works on every occasion. The view does not account for the additional costs of transaction required to initially draft the conditions of contract and subsequently have it 'priced' by tenderers, as was mentioned in Chapter 2.

Lack of familiarity with DEFCON 2000 will produce a learning curve which requires additional resources to understand before it can be employed competently. It is therefore inevitable that contractors will wish to recover this additional cost by adding a premium to their tenders; especially if they are uncertain about the risk apportionment offered in the terms of the contract. The key question is: will the DEO, or other government departments, offer sufficient repeat business to its contractors to overcome these initial adjustments and justify the additional costs?

What was wrong with Latham?

Space precludes a full analysis of the Latham Review and the principles for a 'modern contract'. However Sir Michael Latham identified the duty to deal fairly with each party as an important attribute of a modern contract for construction. DEFCON 2000 was written at the same time as Latham's review and was based on the same lack of satisfaction with the national construction industry. Despite the DEO's rejection of the NEC, DEFCON 2000 compares well with Latham's guiding principles (see Figure 11.2).

That Latham's guiding principles for a modern contract should be the yardstick of good practice remains open to considerable debate. Some of his recommendations (such as the efficacy of trust funds) continue to be contested and unimplemented. Moreover it is questionable whether it is appropriate to stipulate 'generic' guiding principles for every type of business transaction in the construction industry; a theme that is central to this text. Surely a 'horses for courses' approach is more logical depending on the *appropriateness* of a contract's application in the particular circumstances of the works and their business transaction(s). This is not to advocate totally bespoke contracts for every single transaction (for the aforementioned reasons of transaction cost), but surely the application of blanket measures is rather too simplistic to hope for. Thus the DEO is right to re-consider what it wants in a contract, as all clients should; Latham's principles for a modern contract cannot be expected to serve the needs of everyone, nor be accepted by them.

Latham's Principles:	GC/ Works/1	DEFCON 2000	NEC 2nd Edition
Specific duty to deal fairly?	✗	✗	✓
Financial motivation for teamwork?	✓	✓	✓
Package of documents?	✓	✓	✓
Well-defined roles, etc?	✓	✓	✓
Suitable for all projects?	✓	✓	✓
Suitable for all procurement routes?	✗	✗	?
Easy language?	✓	✓	✓
Guidance Notes?	✗	✓	✓
Adjudicator and PM separate?	✓	✓	✓
Choice of risk allocation?	✓	✓	✓
Risk well-allocated?	✓	✓	✓
Advance pricing of variations?	✓	✓	✓
Adjudication on variations?	✓	✓	✓
Payment non-monthly?	✓	✓	✓
Clear payment periods?	✓	✓	✓
Interest on default?	✓	✓	✓
Secure trust funds for payment?	✗	✗	✓
Speedy third party dispute resol.?	✓	✓	✓
Exceptional perf. incentives?	✗	✗	✓
Advance mobilisation pay?	✗	✗	✓

Figure 11.2: Comparing Government Contracts.

The Structure of Power

It is clear that the DEFCON 2000 and NEC families of contracts have been built on opposing philosophies. These philosophies allocate the balance of power between contracting parties in totally different ways. In the following, these approaches are examined to determine which is best in which circumstances.

As outlined in Chapter 8, the NEC is based on trust and mutual co-operation; it believes that close-working relationships are fundamental to successful construction procurement. These relations are expressed up-front in the conditions of contract and established when they are signed. The contract does little to give the client power over the supply base to control the works. Indeed, it is arguable that the NEC philosophy does not believe in a power approach to business transactions. Nevertheless the contract, by its nature, allocates power and risk between the parties and this allocation tends to favour firstly the professional advisers and secondly the contractor. Should the parties to the contract behave opportunistically (i.e. in a totally self-interested manner) it is these parties who, under the the contract, are likely to benefit from the inherent power structure. As contended in Chapter 8, without the presence of an over-determining relationship of mutual trust and co-operation, the NEC structures power against the client's interests and is unlikley to procure construction works effectively. It would be an inappropriate choice for once-off arms-length trading.

The DEFCON 2000 contracts uphold the antithesis of this philosophy. They believe in arms-length business relationships which favour strict liability and absolute compliance with the client's requirements. Trust and co-operation are not requisites to this transactional behaviour. The contracts are very clear in their allocation of power and risk. Control over the works is to be maintained by the client and the risks will be apportioned in an unequivocal manner. Where arms-length relations prevail, there is a *power relationship* in the client's favour. Should the parties to the contract behave opportunistically, the client has the power of redress written into the contract.

Clearly this suggests that DEFCON 2000 is an inappropriate choice of contract for closer working relationships. Indeed where there is the presence of preferential supply relationships, strict adherence to DEFCON 2000 is unlikely to procure construction works effectively.

The question then follows: of the two contractual philosophies, which is the better? The answer, in the keeping with the rest of this text, is that it depends. Indeed neither philosophy is intrisically good or bad. Either approach may be the most appropriate method of procuring construction works effectively; it depends on the nature of the circumstances surrounding the contract.

From the perspective of the client, power and control is the most desirable factor. This gives the client the freedom of choice whether or not to exercise that power and, if so, how. From this discussion, it has been demonstrated that neither DEFCON 2000 or the NEC, is able to give the client this power under all circumstances; there is an over-determining relational influence which presides.

Part E of this book begins to explore the nature of the relationship and its bearing on commercial transactions. The aim of the discussion is to identify under which circumstances the standard forms of contract are most appropriate to be used. From this, we proceed to suggest how the link between the relationship and the conditions of contract can and should be forged and how this marriage can be implemented optimally in the construction industry.

Summary

To summarise, the government contracts (DEFCON 2000 and GC/Works/1) have presented robust alternatives to the NEC and its kin. The NEC considers fairness should be expressed as a requirement to contract in mutual trust and co-operation using a simple form of the English language. However, the government contracts do not put their hope in this or in any other form of 'win-win' outcome. Instead they believe fairness is an expression of clarity, for all to see.

The government contracts are both project orientated, too, but rather than stipulate highly-prescribed management procedures, they choose to leave the supplier to act as it sees fit using its own expertise. Both the GC/Works/1 and DEFCON 2000 suggest that this should be conducted in arms-length manner rather than any collaborative joint-working methods. Their hope is that the parties to the contract may remain focused on the performance objectives of their works rather than becoming entangled in the bureaucracy of contract administration procedures.

Despite the MOD's good intentions to introduce a better contract in the DEFCON 2000, it is unlikely that much enthusiasm will be raised for this 'innovation'. This chapter has expressed several doubts regarding the contract's contribution to the industry and, in conclusion, given the high level of regular spend in government in general (and the defence sector specifically), it is considered that DEFCON 2000 simply reinforces the current blight of arms-length adversarial behaviour that militates against effective construction procurement and provides little basis for clients to understand in a more sophisticated way that which they are seeking to purchase.

Chapter Notes:

1. Powell-Smith V. (1990) *GC/Works/1 Edition 3*, Blackwell Scientific Publications, Oxford, p. 6.
2. Levene P. (1995) *Construction Procurement by Government: An Efficiency Unit scrutiny*, HMSO, London, paragraph 28, p. 23.
3. Latham M. (1994) *Constructing the Team: Final Report of the Government/Industry Review of Procurement and Contractual Arrangements in the UK Construction Industry*, HMSO, London.
4. *New Civil Engineer*, 21 September 1995.
5. This research was facilitated by sponsorship from Railtrack plc and London Underground Limited in the years 1995 to 1997 to determine the optimal forms of contract and dispute resolution methods which exist in the UK construction industry. The authors gratefully acknowledge both this support and the assistance of DEO staff in examining the DEFCON 2000 contracts.
6. The ideas in this section were first published in Cox A. & I. Thompson (1997) 'DEFCON 2000: The MOD's Rebuff of Collaboration' *The International Construction Law Review*, Vol. 14, Part 3 (July), pp. 327 - 341.

7. Cox A. & M. Townsend (1998) *Strategic Procurement in Construction*, Thomas Telford, London.
8. Levene P. (1995) *Construction Procurement by Government: An Efficiency Unit scrutiny*, London, HMSO.
9. Potts K. (1995) *Major Construction Works - Contractual & Financial Management*, Longman, Harlow, p. 101.

Part D

Disputes and Conflict

Chapter 12

Adversarialism and the Escalation of Conflict

Introduction

It can be argued that, if a large part of construction procurement will continue to operate using contracts that attempt to assign roles, rights and responsibilities, then disputes will continue to be an inevitable part of the construction process. The Master of the Rolls, Lord Donaldson has written:

"It may be that, as a Judge, I have a distorted view of some aspects of life, but I cannot imagine a civil engineering contract, particularly one of any size, which did not give rise to some disputes. This is not to the discredit of either party to the contract. It is simply the nature of the beast."[1]

This view is supported by an increasingly popular adage from construction lawyers that:

Construction = Risk = Dispute

It is certain that, as discussed in previous chapters, the construction industry has an established and somewhat tarnished reputation for its adversarialism and levels of dispute and that this cannot be avoided in all circumstances. What is therefore important is knowing when conflict occurs and what can and cannot be done to avoid it.

This chapter serves as an introduction to construction disputes. It determines the extent of conflict inherent in the industry and defines the terminology that has been adopted. The remaining two chapters in this Part examine the existing approaches to dispute resolution and dispute avoidance. These generally comprise well-documented concepts and practices; the industry has considerable focus on this area and these chapters do not attempt to break 'new' ground. Instead, they synthesise the generally agreed 'better practices' to offer a guide to managing disputes.

The last chapter draws on this synthesis in order to develop a conceptual framework that allows parties to construction contracts to consider the issues of conflict, how any conflict is likely to commence and how it can be most appropriately managed. The chapter contends that this area currently lacks theoretical underpinning and robust empirically-verified concepts. It proceeds to offer a conceptual model of conflict and its optimum management in construction, by challenging the current vogue for thinking that conflict should be avoided and/or minimised in every circumstance. This last chapter concludes by considering empirical evidence gathered by the authors to verify/falsify this thinking.

Background

Many leading industry reviews and reports have concurred that the construction industry is adversarial and that this behaviour has a detrimental impact on performance. However there has been very little work to quantify how much waste and inefficiency is caused by these prevailing attitudes.

One attempt to quantify the level of conflict was made in a survey in the *Financial Times* (12 October 1995). It revealed that approximately one-quarter of contracting firms earned between 10% and 15% of their annual turnovers from contractual claims against customers and suppliers, while a further third earned between 5% and 10% from such claims. Although the total cost of these claims and disputes was not reported, a reasonable calculation can be made using available data sources. According to published listings by the *New Civil Engineer*, the sum total of the top-50 UK contractors' annual turnovers was £21,293 million in 1995/96 which, by averaging out the *Financial Times'* statistics, suggests the total sum of these claims costs the industry approximately £1200 million *per annum*.

According to Department of the Environment statistics, annual expenditure in the construction industry is £49.8 billion[2], which means that the top-50 contractors have a 43% market share. It is reasonable, therefore, to extrapolate the total annual cost of conflict by this proportion (assuming the level of conflict is relatively constant throughout the industry), which suggests an annual cost of £2800 million (i.e. 5.6% national construction expenditure).

Although this expenditure has been 'justified' in the form of agreed and settled claims or disputes (since it is contractors' income), it still represents a considerable proportion of *additional* expenditure for clients which had not been considered at the contract-award stage. Furthermore, not all claims and disputes are settled ('won') and so this figure does not reflect the true degree of claimsmanship or adversarialism, which will be considerably greater than this figure. Research from arbitration cases in the early 1990s[3] has suggested that small claims (i.e. up to £2500) are settled at an average amount of 65% of that claimed, whereas claims in excess of £40,000 are settled at approximately 40% of that claimed. Initial claims and their subsequent disputes are clearly inflated beyond their settlement sum. By making an assumption that the 'inflated-claim' rate for arbitration cases can be extrapolated across all claims and disputes, it is reasonable to suggest that the industry disputes over 10% of its business at any given time (i.e. £5 billion p.a.).

However, this figure only represents the level of conflict at the first 'tier' in the construction supply chain: the main contractors. It does not consider the subcontractors and materials suppliers further upstream in the supply-base where the number of construction firms and degree of fragmentation are greater; nor does it consider the potential conflict that exists between the client and its professional advisers. Thus these estimated figures are very conservative and in reality are likely to be considerably higher.

Furthermore, the Chairman of the Association of Regulated Procurement (a construction lawyer) recently estimated that the construction industry spends at least £500 million *per annum* on professional/legal fees either in support or defence of contractual disputes[4].

Clearly if the industry was less adversarial and claims-conscious some, if not most, of this additional expenditure could be saved, as well as the associated internal on-costs of managing the conflict. The effect would be an immediate saving on construction costs.

It follows that the construction firm in the supply/value chains which can manage its business relationships optimally is able to appropriate/accumulate these savings for itself. When this is set in the context that most major contracting firms are only receiving an operating profit of between 1% and 2% of their annual turnovers, the importance of this additional revenue stream cannot be overlooked or understated.

Thus the appropriate management of relationships and avoidance of *unnecessary* dispute provides an opportunity for construction firms to gain significant [sustainable] competitive advantage over those who do not take this approach. It is worth stressing, however, that adversarialism is inherent in many areas of construction and cannot be avoided by everyone.

How should we respond?

Given the costs of conflict, two differing views can be developed:

1. Disputes are wasteful of a firms' resources and therefore should be avoided, wherever possible. Where this is not possible, disputes must be minimised and any that do occur must be resolved as *efficiently* as possible.
2. Conflict is inevitable in business relationships and must be managed appropriately. Thus, under certain circumstances disputes should be avoided, whereas under other circumstances, dispute avoidance is undesirable. This view considers that existing disputes should be resolved as *effectively* as possible, which may mean an adversarial pursuit of conflict in order to maximise business objectives.

In Chapter 14 we aim to distinguish between these two views. The principal differences lie in foundational business philosophies. The 'efficient' school of thought considers that conflict is 'wrong' and therefore dispute avoidance and/or minimisation should be a goal/pursuit of all firms in construction. As discussed at the end of Chapter 11, the NEC falls into this school. Conversely the 'effective' school of thought asserts that all business is conflictual as it is the pursuit of an organisation's individual business goals that is of primary importance. Conflict will always be present in [external] commercial transactions since contracting firms' individual business goals often differ. The consequence of this view suggests that a firm should manage conflict in the most effective way to achieve its end-goals.

In considering these schools of thought and the supporting literature on the subject, a conceptual framework for the management of conflict can be established. This has been developed in Chapter 14 where both schools of thought are reviewed and challenged.

The majority of literature in construction management considers that all disputes should be avoided/minimised wherever possible (a view typified by the Latham Review). Similarly, much of the literature in purchasing and supply chain management supports this view (e.g. the 'lean' concept). However this view should be questioned as there appears to be some confusion between *means* and *end*. The notion that conflict must be avoided as an 'end-state' is myopic as it will not necessarily achieve business objectives. The end-goal should be the maximisation of business objectives; conflict minimisation or conflict propagation are simply alternative means to this end and must be considered appropriately in the context of the governing transactional relationship.

Definitions

Before a discussion on disputes can be meaningfully presented, it is important to ensure terms are clearly understood and defined.

The following chapters are examining *inter-organisational* conflict; that is, a firm's conflict with external parties. Internal conflict within the organisation has been examined by several distinguished writers[6] but few have extended this study to conflict beyond the boundary of the firm. This Part examines the conceptual basis for conflict between two separate business entities to a commercial transaction.

There is an escalation process involved in the development of conflict (see Figure 12.1). This illustration is helpful to consider the process by which conflict develops. At each stage there is a breakdown, or failure, in the *interests* of the parties and conflict escalates. The process begins with a problem or issue, in which there lies a difference of opinion or views between the parties over some material value. Unless that difference is resolved, disagreement ensues and conflict is established. Since the formation of a contract requires agreement as one of its fundamental tenets, it follows that conflict relates to matters which are considered (by at least one party) to be beyond the scope of the contract and upon which agreement cannot be made.

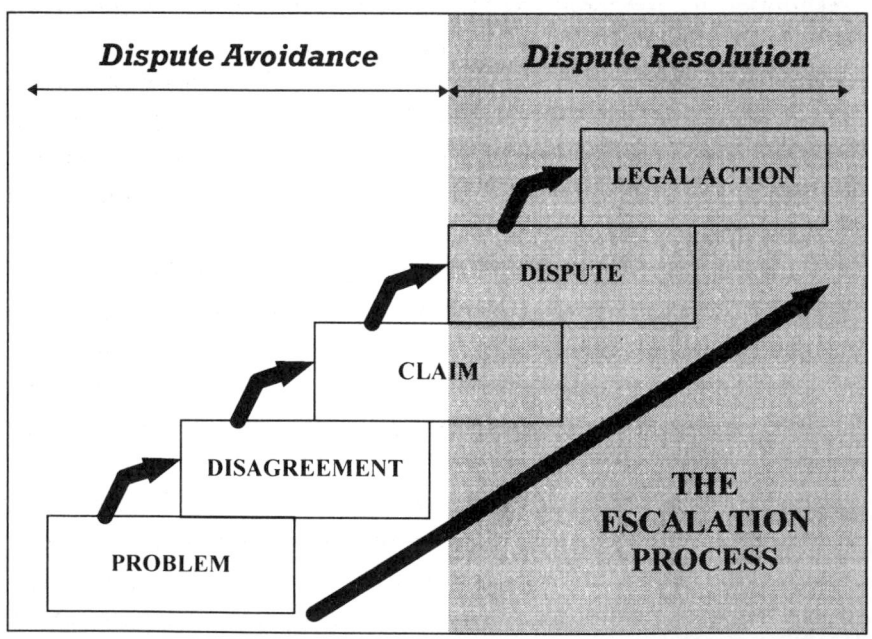

Figure 12.1: The Conflict Escalation Process.

The recognition of the disagreement is formalised upon the submission of a 'claim' from one party to the other. Claims have material consequences and usually comprise the submission of written details requesting extra-contractual consideration (i.e. additional allowances for payment and/or time). It is at this stage in the conflict process that disputes will either occur or be prevented. Every claim demands a decision on its validity which the claimant has an opportunity to accept or reject; to reject a decision is to dispute its merit and this is the starting place of all disputes (i.e. there must be a decision to dispute).

However in practice the process and distinction between these stages is not as precise as this conceptual approach suggests. Claims and counter-claims occur between the parties as part of general negotiation positioning over an issue or problem. That is not to say a dispute has occurred or is necessarily likely to occur; a dispute is only formed when the parties' positions on an on-going claim are irreconcilable and the disagreement becomes entrenched. Thus in practice there is a lack of clarity concerning the origin of disputes; their boundary with claims is blurred and generally ill-defined throughout the construction industry.

Formally, a dispute is said to have been lodged or notified upon the issue of a 'Notice of Dispute' from one party to the other, or to the presiding third party professional [Engineer or Project Manager], in accordance with the contractual conditions[7]. Of course, this is purely a bureaucratic origin; the true inception of a dispute will be long before the issue of such a Notice, when the parties begin to contest a problem/claim within the progress of the works. Despite one commentary stating: *"...no notice, no dispute..."*[8], we contend that it is naive to assume that the presence of a Notice will be the determining factor as to whether or not a dispute exists.

Thus, for the purposes of this book, a **dispute** shall be taken to be *"a contested difference of opinion between contracting parties which carries commercial consequences"*. Usually these consequences are limited to issues of cost, time or quality in the work particulars; but they can also include more fundamental matters associated with the conditions of contract.

Similarly **conflict** shall be considered to be an umbrella term regarding *"the failure of two contracting entities to agree issues associated with the performance of a commercial* *transaction(s)"*. Thus conflict, when manifested, may comprise disagreements, contractual claims and/or disputes, as already described.

Chapter Notes

1. See Foreword of Hawker G, J. Uff & C. Timms (1986) *The Institution of Civil Engineers' Arbitration Practice*, Thomas Telford, London.
2. DOE Construction Industry Board (1997) *The State of the Construction Industry*, Issue 7, (February).
3. Bloore R.D.S. (1995) 'Flip-Flop Costs – A tonic to revive arbitration?' *Arbitration*, Vol. 10, No. 2.
4. Clark B. (1996) 'Dispute-free Contracting' Presentation at 'Taking Latham Forward through Procurement-led Initiatives' CPN Workshop, Edgbaston, Birmingham, 25 September.
5. For further discussion on the lean supply concept see Womack J.P. & D.T. Jones (1996) *Lean Thinking*, Simon & Schuster, New York. For a counter view see Cox A. (1997) *Business Success: A Way of Thinking about Strategy, Critical Supply Chain Assets and Operational Best Practices*, Earlsgate Press, Boston.
6. See for example: Fox A. (1966) *Industrial Sociology and Industrial Relations*, HMSO, London; Pascale R.T. (1990) *Managing on the Edge: How successful companies use conflict to stay ahead*, Penguin Books, London; Handy C. (1993) *Understanding Organisations, 4th Edition*, Penguin Books, New York; Daft R. L. (1995) *Organisational Theory & Design, 5th Edition*, West Publishing Company, St Paul, MN.
7. See for example Clause 66 of the ICE Conditions of Contract 6th Edition.
8. Cottam G. & G. Hawker (1991) *ICE Conditions of Contract for Minor Works: A User's Guide and Commentary*, Thomas Telford, London.

Chapter 13

Dispute Resolution

Introduction

The previous chapter suggested that escalated conflict is manifested in the contractual dispute. This chapter examines current dispute resolution practices and asks how appropriate each one is in providing the longer-term business objectives of the parties. Where appropriate, reference to supporting literature is made for readers who wish to read more on the subject.

Dispute resolution procedures refer to the methods used in practice to settle a dispute between two parties within a contract. As already noted in Chapter 12, this does not include dispute prevention or avoidance (these are covered in Chapter 14), as such, these are measures to be employed *ex post* the manifestation of a dispute (i.e. in a reactive manner).

For classification purposes we have identified four common categories of dispute resolution method:

- Resolution by Agreement;
- Resolution by Intervention;
- Resolution by Third-Party Determination;
- Resolution through Litigation.

In the following, each category has been examined in detail. Towards the end of this chapter, we begin to demonstrate how these dispute resolution techniques can be successfully combined in order to effectively resolve contested matters.

Resolution by Agreement

Resolution by agreement is perhaps the most regularly practised form of dispute resolution and yet least recognised (very few texts recognise this as a method of dispute resolution for construction). It is the least formal, least time-consuming, least expensive, and yet undoubtedly the most successful of methods. This is so even to the point that many fail to consider it is part of the dispute resolution process. However we suggest that an attempt to settle the dispute through negotiation and agreement should be the starting place of all disputes.

There is no defined means of reaching a settlement this way, other than simply by discussion and agreement between the parties to the contract: it is simply *negotiation* until a point of agreement has been reached. Regrettably, parties are often too quick to entrench themselves into their 'sides' of an argument, which then becomes a personal agenda belonging to the individuals concerned.

The negotiations may be carried out at a relatively junior level on site, or at the most executive inter-organisational levels; although rarely at the same time. Often there will be a series of negotiations at increasingly higher levels in the two organisations. The procedures could be largely informal and haphazard (in the form of haggling a barter) or more strictly formal and 'scientific' (as in the form of claims management and assessment). Either way the process involves sufficient interaction between the parties to accomplish a settlement.

One outcome may be a decision to refer the dispute up the line of command to more senior management. More formal negotiation procedures will allow this form of 'structured' negotiation in order to keep the dispute 'fresh' and prevent stalemates occurring between the parties. The option to refer also acts as an incentive to ensure the dispute is legitimate and well-documented (as it has to be presented to one's seniors for them to defend/pursue). This also ensures the dispute is referred to the appropriate level of authority within the organisation to make a decision. Senior management is less involved in the day-to-day detail and can therefore offer a more objective view on contested matters; it should comprise the more experienced members of the industry with a better understanding of the overall business processes/drivers of their organisation.

Structured negotiation has been introduced by a few forward-looking major clients with great success (see Case Study 9). However it seems that few firms have yet to implement this simple operational approach. In one of our research surveys conducted in early 1997, the top-100 contractors and the top-100 construction clients were asked whether they employed formal negotiation procedures. Despite the fact that only 24% of the respondents said that they did, it was clear from other responses that those who employed these procedures enjoyed higher rates of success at resolving disputes by negotiation.

There are numerous books and manuscripts on the art of negotiation and a repeat of their contents is unnecessary, except to state that the skills required to conduct successful negotiations should be part of the core competencies of any procurement and/or commercial professional. These texts predominantly comprise operational tactics and have very little strategic overview and few make any link between negotiation and dispute resolution procedures[1].

A final word needs to be given to the role of the Engineer to the contract. Specifically in the ICE 6th Edition, but also more generally to all forms of contract which employ a professional third party in a quasi-arbitral role. The position of Engineer is often given to a senior member of the client organisation (or their representative's organisation) and under certain conditions he is asked to preside over specific decisions at various points in the performance of the contract. This can create a conflict with the mechanism of structured negotiation; particularly if a contested matter is referred to senior management where the Engineer is employed. Referral of the dispute is not likely to produce a satisfactory outcome if the Engineer has already been involved in the matter(s) at an earlier time in the proceedings. It is unlikely (but not inconceivable) that someone would review the same matter twice with different outcomes and thus, in this event, the negotiation mechanism could break down and protract the dispute. If a structured negotiation clause is to be written into a contract, then this issue must be borne in mind. The Engineer should not be part of process in more than one place (especially if he is later required to give an 'Engineer's decision' as in clause 66(3) of the ICE 6th Edition).

Case Study 9: Successful Dispute Resolution at Railtrack plc

The 35 Railtrack RT/1a maintenance contracts engage sole suppliers (the former infrastructure maintenance units) to carry out all Railtrack's infrastructure maintenance for 5, 6 or 7 years from 1994, at a value of approximately £585 million *per annum*. In order to ensure successful performance, the contract needs to allow a governing relationship to exist between Railtrack and its suppliers. The contract has a structured procedure for resolving disputes which comprises negotiation followed by third party adjudication (clause 28). The clause stipulates that for any dispute arising, the parties must endeavour to resolve it by discussions extending over at least one month at Director level, before the matter is referred to the adjudicator. Evidence from the LNE Zone suggests that 70–80% of disputes are settled at project management level, and of the outstanding disputes *c*. 90% are resolved at Contract Review stage with the remaining negotiated at senior management level. After three years of the RT/1a contracts, no disputes anywhere within Railtrack had made it beyond negotiation to adjudication.

The merits and benefits of negotiated settlements rest with the pragmatism they offer. For those who wish for an 'exact' result, and are not prepared to come to a quick convenient compromise, these same merits will be seen as detriments. The following charts summarise the merits, potential pitfalls and critical success factors for resolution by agreeement.

Merits of Resolution by Agreement

- quick;
- inexpensive;
- informal;
- does not consume large quantity of internal resources to conduct;
- confidential;
- non-disruptive to the rest of the works, or other business contracts;
- can be managed within the context of the business relationship;
- can be carried out at arms-length or collaboratively (as appropriate);
- rarely harms prospective (future) business relationships;
- 'commercial' settlements/trade offs can be agreed.

Potential Pitfalls of Resolution by Agreement

- personalities of individuals concerned;
- entrenched views;
- behaviour which is not appropriate to the business relationship;
- vested interest from management (or individuals);
- short-term financial drivers (e.g. quarterly performance reports);
- inappropriate, insufficient and/or inaccurate information;
- poorly presented claims/disputes;
- inappropriate use of power in the business transaction;
- lack of delegated authority to make critical decisions;
- difficult to justify accountability to external audit teams;
- over-emphasis on 'strict liability' (i.e. the other side is 100% wrong!);
- prolongation of the dispute.

Critical Success Factors of Resolution by Agreement

- appropriate relationship between the parties (see Chapter 15);
- sufficient interaction to allow a settlement to be reached;
- accurate information to the right level of detail;
- clear presentation of the case (cause and effect);
- competent staff (on both sides);
- authority to reach settlement;
- willingness to compromise (to a degree);
- business acumen.

Resolution by Intervention

Traditionally, when the parties to a dispute reached an impasse in their discussions, or when negotiations failed and the two 'sides' became entrenched, the only way forward was to either go to court or proceed with an arbitration on the issue. However the advent of *Alternative Dispute Resolution* (ADR), first in the US and more latterly in the UK, has recognised that there are intermediary steps between negotiation and arbitration/litigation. The international organisation FIDIC refers to these steps as 'amicable settlement of disputes', as the parties have the option to agree a settlement between themselves before taking legal action[2]. This section examines the mechanisms of *mediation* and *conciliation* as natural extensions of the negotiation process by employing a neutral third party to intervene and facilitate discussion and resolution (the processes have also been called 'assisted negotiation').

Mediation and conciliation are often confused with each other; however they are not the same. The difference lies in the extent to which the third party intervenes with evaluative statements: the mediator merely facilitates discussion, whereas the conciliator provides the parties with a recommendation based on his opinions of the dispute.

Mediation has been defined as *"A without prejudice non-binding dispute resolution process in which an independent third party ("neutral") assists the parties to settle their differences, but does not advise them of his own opinion as to the issues and merits of the dispute"*[3]. In 1990 the CBI launched the Centre for Dispute Resolution (CEDR) as a non-profit-making organisation to promote and encourage mediation and other cost-effective approaches to dispute handling. Since its establishment, it has an impressive record of resolving 95% of all disputes referred to it.

The process of mediation has been likened to Henry Kissinger's 'shuttle diplomacy': the parties meet in joint and separate private meetings (called 'caucuses') which last for up to two days during which the mediator moves between them attempting to reach a settlement. Because of the speed of the process, the associated costs are negligible; the parties are expected to pay for all costs (such as room hire, etc.) and the mediation fees.

Although in pure terms the mediator does not make evaluative statements, it is arguable that the nature of questioning by the mediator can be 'evaluative'. Most mediators will resist any such angles to their questioning or statements as it can be seen to jeopardise their neutrality and ultimately the efficacy of the mediation. There are other forms of evaluative mediation (as noted later), but these are not as commonly used.

It is to be stressed that mediation is a totally voluntary exercise, which once started does not have to be fully carried out if either party wishes to pull out. To be employed it does not need clauses within the original form of contract, although CEDR has published its own 'model' clauses for clients who wish to pursue mediation as a potential dispute resolution process in their contracts. The inclusion of 'alternative' dispute resolution clauses in a contract can place an unintended obligation on the parties to follow the prescribed method (whether they wish to or not). In holding that this pre-agreed method should be followed, Lord Mustill has stated: *"...those who make agreements for the resolution of disputes must show good reasons for departing from them.....The fact that the [employers] now find their chosen method too slow to suit their purposes, is to my way of thinking, quite beside the point."*[4]

However it is unlikely that the courts would enforce a non-binding negotiation method (such as mediation) on the parties. The most likely determinant of a successful resolution will be the *intention* of the parties to agree to mediate (as neither party can be forced). Similarly timing will be essential since mediation will only succeed if the parties are willing to compromise. Furthermore, the parties do not have to accept the settlement, but it is usual that once an outcome has been agreed it is documented and signed as a contractually binding agreement. This *will* be enforced by the courts.

The voluntary nature of mediation has both advantages and disadvantages; it can be difficult to get the disputing parties to agree on this method of resolution (their agreement may depend on how strong they perceive their persuasions to be and they may prefer to chance their fortunes on other resolution practices, or just hold out in the hope the other party caves in), however once agreement has been reached the parties are said to have 'ownership' in the settlement and this can often help preserve business relationships.

In the United States, Florida State has introduced *compulsory* mediation in an attempt to resolve disputes. It is curious how parties can be expected to agree a settlement between themselves against their wishes, and it is unlikely that English law will follow. However, following a High Court Practice Direction in 1995, all law firms must now offer ADR as part of their services. At the time Lord Chief Justice Cresswell announced that judges would not be acting as mediators, or become involved in any ADR processes. However his successor, Mr Justice Waller, has recently announced that judges have an option to adjourn court proceedings in an attempt to 'encourage and enable' the parties to find a settlement through ADR. This has been referred to as 'early evaluation' where the judge invites parties to accept a recommendation based on his initial impression of the disputed case and thus prevent an expensive and prolonged court case proceeding.

Such evaluative approaches to mediation are more akin to the conciliation process. The role of the conciliator is *"...to move backwards and forwards between the parties explaining the point of view of each other to the other"*[5]. The most common use of conciliation in construction is the Institution of Civil Engineers' Conciliation Procedure. This procedure is built into the ICE contracts and the FCEC form of subcontract; it has been accepted by the Institution of Civil Engineers, the Association of Consulting Engineers and the Federation of Civil Engineering Contractors. It should also be noted that it is very similar to the UNCITRAL conciliation procedure.

As with mediation, the conciliation process is optional and is usually entered into through an agreement to employ the Conciliator and pay his fees. The process is confidential and the parties are required to indemnify the conciliator against subsequent legal action. The parties prepare statements of their view of the disputed issue to send to the conciliator. The conciliator then begins an investigation into the matters of dispute, by interviewing the parties and/or visiting the site. Unlike an arbitrator, the conciliator does not have to observe the laws of natural justice and is given inquisitorial powers to elicit any omitted information which may assist the investigation.

Having established a full understanding of the issues, the conciliator meets with the parties (sometimes in the presence of an 'expert' contracted in to advise on technical issues) to begin to form an opinion. The conciliator discusses the merits of the party's case with each of them, exploring both strengths and weaknesses of their arguments and offering an opinion of how an arbitrator may view the dispute. If during these exchanges the parties can be persuaded to compromise and reach a settlement then the process has been successful and there is no need to pursue the conciliation to its ultimate end. This is an argument that has been put forward in defence of conciliation over and above adjudication. The argument suggests that conciliation is better because it attempts to get the parties to agree a settlement primarily and, failing that, offers an expert's opinion on the matter (which does not have to be accepted if one party thinks it knows better).

If agreement cannot be reached in the course of the process, the conciliator is required to fulfil his duties to the Conciliation Agreement and issue his own recommendation for the settlement. It should be noted that the conciliator's recommendation is:

- only the *opinion* of the conciliator;
- not justified with reasons (unless expressly required, in which case these will be documented separately);
- not binding (i.e. 'without prejudice');
- not necessarily compliant with law or statute (it will usually provide a commercial solution with pragmatic intent);
- usually only offered after his payment has been made.

In total, the process takes two months to form a recommendation followed by upto one more month to conclude the process.

The merits of 'resolution by intervention' are similar to those for direct negotiation, inasmuch as the dispute is often resolved at a 'reasonable' compromise position between the parties. The advantage is that, in bringing in a neutral third party to intervene, the parties perceive their dispute has been justified by an outsider and not just brushed aside by the other party.

Merits of Mediation / Conciliation

- speed (for mediation just a few days, for conciliation not exceeding 2–3 months);
- inexpensive (usually a factor of the time involved and a fraction of the disputed sums, i.e. sometimes less than 1%);
- does not consume large quantity of internal resources to conduct;
- unlike internal negotiations it is easy to audit;
- confidential;
- non-binding;
- parties are able to agree settlement;
- relatively objective;
- 'commercial' settlements/trade-offs can be agreed;
- potential to restore business relationships.

Potential Pitfalls of Mediation / Conciliation

- inappropriate, insufficient or inaccurate information;
- poorly presented claims/disputes;
- failure to agree on who the mediator/conciliator should be;
- inability to compromise;
- lack of trust in mediator's/conciliator's experience, ability and/or neutrality;
- fear that concession at mediation will lead to exposure in any subsequent arbitration.

Critical Success Factors in Mediation / Conciliation

- an intention in the parties to have a settlement reached;
- clear presentation of the particulars of the dispute (cause and effect);
- accurate information to the right level of detail;
- ability to see strengths and weaknesses of both sides of the dispute;
- competent staff within the parties;
- a competent mediator/conciliator;
- the neutrality of the mediator/conciliator;
- the parties' trust in the mediator's/conciliator's experience, ability and neutrality;
- willingness to compromise (to a degree);
- business acumen.

Third-Party Determination

As a dispute becomes more protracted, the parties generally lose the ability to accept the other's point of view as their own becomes increasingly entrenched. The likelihood of amicable settlement and/or compromise decreases. If the parties have attempted to resolve the dispute through facilitated negotiation and have failed, it is unlikely the parties will ever be able to reach a settlement by agreement.

The 'next step' requires a third party presiding on the issues and pronouncing a judgement of his/her own, in order to prevent the dispute getting into the courts. It is at this point that the resolution processes begin to assume a life of its own and the parties lose control on the outcome; others are deciding how the dispute will be settled.

Adjudication and arbitration are the two dispute resolution processes discussed in this section. Although their mechanisms are very different, their principles are in essence quite similar: the parties agree who shall play the judicial role, present their cases and receive a judgement. However, advocates of either procedure would be quick to point out the vast differences between the two processes and, because of these, they should be considered apart from each other.

The Oxford English Dictionary defines an adjudicator as one who settles a question and thus is one who performs a judicial *function* (even if not all the hallmarks of judicial *practice* are present). There is still widespread confusion in the industry between adjudication and arbitration and to date little has been published on adjudication. One of the main differences is that arbitration is enshrined in statutory law and has specific [essential] characteristics which define it (see Figure 13.1). This statutory underpinning means that even if a contract refers a dispute to adjudication (by name) but the guiding principles of that process are essentially that of arbitration, the process is in fact an arbitration governed by the Arbitration Acts. The use of the word 'adjudication' is not decisive on the resolution process to be adopted. In addressing this particular point, His Honour Judge Humphrey Lloyd QC has said: "...*It is plain that 'adjudication' taken by itself means a process by which a dispute is resolved in a judicial manner. It is equally clear that 'adjudication' has as yet no settled special meaning in the construction industry...* "[6].

With this in mind, it is therefore important to consider:

- what are the differences between adjudication and arbitration?
- why is there a need for an additional process when arbitration is already recognised and long-established?
- and, principally, what are the relative merits of each process?

ESSENTIAL ATTRIBUTES OF ARBITRATION

- The agreement pursuant to which the process is, or is to be, carried on ("the procedural agreement") must contemplate that the tribunal which carries on the process will make a **decision which is binding on the parties** to the procedural agreement.
- The procedural agreement must contemplate that the process will be carried on between those parties whose substantive rights are determined by the tribunal.
- The jurisdiction of the tribunal to carry on the process and to decide the rights of the parties must derive either from the consent of the parties, or from an order of the court or from a statute the terms of which make it clear that the process is to be arbitration.
- The **tribunal must be chosen**, either by the parties, or by a method to which they have consented.
- The procedural agreement must contemplate that the tribunal will determine the rights of the parties in an **impartial** manner, with the tribunal owing equal obligation of fairness towards both sides.
- The agreement of the parties to refer their disputes to the decision of the tribunal must be intended to be **enforceable in law**.
- The procedural agreement must contemplate a process whereby the tribunal will make a decision upon a dispute which is already formulated at the time when the tribunal is appointed.

Figure 13.1: The essential attributes of arbitration.

Source: Mustill and Boyd[7], emphasis added.

Essentially the differences between adjudication and arbitration lie in the way the proceedings are conducted and the nature of the judgement. The adjudicator makes a decision (which can be binding or non-binding[8]) after examining the evidence and making any necessary inquiries; he is an expert using his own skill and experience to judge. Conversely the arbitrator makes an award (which is final and binding) after receiving the presentations and defence of the parties in the course of *natural justice*.

ADJUDICATION:	ARBITRATION:
• quick: 28-day fixed timescale; • [therefore] less expensive! • inquisitorial powers; • adjudicator acts as an 'expert' not an arbitrator; • makes *decision* on skill and experience as well as evidence; • adjudicator can be sued for negligence; • non-binding unless otherwise pre-agreed as a contractual term; • therefore risk that this is not the end of the dispute (i.e. it is 'temporary'); • review by tribunal if required; • loss of natural justice; • no legal representation; • impartial; • cannot adjudicate on points of law; • covered by Housing Grants Construction & Regeneration Act 1996.	• no limit on timescale (open-ended); • no limit on expense; • quasi-judicial with some inquisitorial powers; • arbitrator is not an 'expert'; • makes *award* on evidence of case; • cannot be held in negligence; • final and binding (unless decision was made beyond his jurisdiction or contrary to current law); • courts may enforce award; • full exposure to natural justice; • parties are legally represented; • impartial; • cannot arbitrate on points of law; • covered by Arbitration Acts 1950, 1975, 1979 and 1996.

Figure 13.2: Adjudication and Arbitration compared.

The comparison is best illustrated in Figure 13.2. To the ignorant these differences are small and insignificant; however to the legal fraternity they have a significant impact on the dispute and the efficacy in which it is resolved. These differences are best considered in answer to the aforementioned question: why an additional process was required either before or in place of arbitration/litigation. One suggestion has been that Sir Michael

Latham's preference for adjudication was not based on the advocation of a new practice, but more instinctively he was looking for a new vehicle to recover the lost values of past good arbitral practice[9]. Whether this suggestion of Latham's intentions is an accurate reflection is somewhat irrelevant. The principal concern rests with the nature of arbitration as a commercially pragmatic alternative to the courts and whether its position has been jeopardised by recent practice. If this is the case then adjudication appears to be a welcome innovation.

The use of arbitration as a pragmatic and efficacious alternative to the courts can be considered under the following headings of 'speed and expense', 'finality' and 'natural justice'.

Speed and expense

It has been long recognised that arbitration has become increasingly prolonged and complex. It was originally intended to be a quick and confidential alternative to the 'laboratory of the courts'. However current opinion seems to consider that arbitration now no longer makes 'commercial sense'. Past president of the Chartered Institute of Arbitrators (CIArb), Sir Michael Kerr, has referred to some lengthy arbitrations as a 'forensic obscenity'[10] and a former CIArb Chairman noted that arbitration has been accused of being as time consuming as litigation and even more expensive[11].

Clearly the speed of the process is important especially as expense is a corollary of its duration. There needs to be a balance between the necessity to fully explore the issues of the dispute (factual, technical and legal complexities) and practical realities of allowing an unresolved matter to continue (in terms of both direct and indirect costs). Herein the views of business and law may conflict. But surely commercial pragmatism must prevail: few businesses can afford to take the risk of electing a protracted method of resolution unless certain that it will reap its rewards.

It is not sufficient to suggest the problem lies with arbitration *per se*. Some blame the time and expense involved in arbitration on its having "*...come under the control of the lawyers*". Others believe it is the English adversarial system which has made arbitration more confrontational and that the English legal profession has transformed the once-informal arbitration into 'wigless litigation'[12].

Although the speed and expense argument presented here would seem to heavily favour adjudication over arbitration, there are some issues which counter these arguments. In promoting a quicker resolution, something needs to be sacrificed in terms of the thoroughness of the proceedings. In the case of adjudication, this is legal representation and evidence. Full and proper consideration cannot be given to all the particulars of the dispute, and this compromise needs to tolerated by the parties if they are to accept adjudication. The argument in favour of arbitration is that of the superficial nature of adjudication; but not everyone subscribes to it.

Finality

Given the aforementioned compromise in favour of a quicker and cheaper process, it would be inappropriate for an adjudicated decision to be final and binding. However, in arbitration the settlement is; the arbitrator makes an *award* which is protected by the Arbitration Acts which the courts will uphold. There has been considerable discussion surrounding the question of finality in the adjudicator's decision. The current legal position is that adjudication is non-binding unless otherwise agreed by the parties[13]. The effect of making the decision 'temporarily binding' counters the previous argument of the superficiality of adjudication. It also raises the a 'barrier to entry' against arbitration/litigation: the parties have already received one judgement, do they really wish to pursue an expensive and time-consuming process at the risk of receiving the same outcome?

Natural justice

The essential rules of natural justice are: (1) the right to present one's case to its best advantage; and (2) the right to be told the detail of the other party's case and the opportunity to refute it. Given the timescales of the adjudication procedure (28 days under the Housing Grants Construction & Regeneration Act 1996), these rules are inevitably going to be compromised. The choice facing the parties rests on whether speed with compromise is to be favoured more than exactitude with prolongation.

There are arguments for either approach: arbitration allows the full course of natural justice to be preserved and honoured, but then so do the courts. As the Master of the Rolls has said: *"...the arbitral process, by mimicking the processes of the courts, and becoming over-legalistic and over-lawyered, has betrayed its birthright by allowing itself to become as slow, as expensive and almost as formal as the court proceedings from which it was intended to offer escape."*[14].

In the above considerations, adjudication has been presented as the commercially preferred method of resolving disputes through a confidential out-of-court 'judicial' process. Many consider arbitration to be out-dated, expensive, slow and overtly formal; which are the exact issues it was originally intended to address as a legally-binding alternative to the courts. It would appear that adjudication has successfully stolen the 'higher ground'. For each of these points, the efficacy of resolution is greater with adjudication. Therefore it must be questioned whether there still remains a place for arbitration as an effective method of resolving disputes.

Merits of Adjudication / Aribitration

Adjudication	Arbitration
• quick: 28 days;	• may start at convenient time and take as long as parties need;
• inexpensive;	
• non-binding;	• binding;
• confidential;	• confidential;
• no need for legal representatives to become too involved.	• parties can be legally represented;
	• natural justice;
	• thorough.

Potential Pitfalls of Adjudication / Arbitration

Adjudication:	Arbitration:
• inappropriate, insufficient or inaccurate detail;	• inappropriate, insufficient or inaccurate detail;
• poor presentation of case;	• poor presentation of case;
• dependent on competence of adjudicator;	• dependent on competence of arbitrator;
• justice not seen to be done;	• justice not seen to be done;
• failure to agree on adjudicator;	• failure to agree on arbitrator;
• time taken to appoint adjudicator prolongs dispute.	• parties unable to 'force the pace' of the procedure.

Critical Success Factors of Adjudication / Arbitration

Adjudication	Arbitration
• competent adjudicator;	• competent arbitrator;
• impartiality of the adjudicator;	• impartiality of the arbitrator;
• clear presentation of case;	• clear presentation and defence of case;
• accurate information to the right level of detail;	• competent legal representation;
• acceptance of the adjudicator's decision.	• accurate information to the right level of detail.

Resolution through Litigation

Generally, any dispute may be taken to the courts to be resolved unless there is an arbitration clause in the contract documentation which, as already agreed by the parties, must be carried out thus denying the right to prosecute. This process is called litigation. It forms the 'final say' on the disputed issue (save for an appeal) but it is usually hoped that the issue will be resolved before the process is fully instigated. Usually just the threat of the ensuing legal battle will speed disputing parties to agree a settlement 'on the steps of the court room'.

There are different courts in which a case may be heard depending on the size or nature (and sometimes location) of the dispute. However most construction cases are normally heard in the Queen's Bench Division of the High Court. The hearing is usually in front of one judge (known as an 'Official Referee'), sometimes assisted by 'assessors' who offer technical guidance on the particulars of the case. Judges are used to hearing building and civil engineering disputes, as approximately 80% of the Official Referees' cases are construction-based [15].

Although litigation is usually considered to be the last say on a disputed matter, dissatisfied parties may challenge the verdict by taking it to the Court of Appeal, or even further to the House of Lords under certain circumstances. If this course is taken it can be many years before the dispute is eventually settled.

As may be expected, the trial is adversarial and the judgement is final and binding on the parties. In the judgement, liability is decided and any monetary awards are made; these will usually comprise any monies due plus the winner's costs of the resolution procedure. It is not uncommon for the legal costs to be in excess of the award. For example, the Abbeystead disaster of May 1985 (settled in 1988) awarded £2.2 million compensation to the plaintiffs, with a further £3 million in legal fees.

The recent Woolf report (described by himself as *"...a new landscape for civil justice"*) has introduced a three 'track' system to the courts in order to make the process a little more affordable. The first track is for disputes upto £3000 in value which are taken through the small claims court (which is already in existence). The next track (the 'fast track') is for cases upto a £10 000 limit for which there is a timetable restriction of 30 weeks for the procedure culminating in a maximum 3 hour trial. In these trials all evidence is to be provided in writing, cross-examinations are dispensed with and the judge will spend the time reviewing the evidence and questioning witnesses.

Lord Woolf's final track for litigation is the 'Multi-track' which is for complex cases in excess of £10,000. Judges will set a timetable for the proceedings and then manage it on a programmed basis. It is hoped that the effect of these reforms will make the courts more accessible and commercially viable than has been the case before. There is a clear change in emphasis, moving away from prolonged overtly-legalistic hearings, towards quicker less expensive proceedings. The market-place has finally arrived on the doorsteps of the courtroom and these changes must be broadly embraced throughout the industry. They will make for easier and more cost-effective procedures. Perhaps the only reservations will come from the legal fraternity as they see their workload diminish. For those who regularly conduct business in construction, this may not be too bad a thing!

Merits of Litigation

- full access to natural justice;
- full legal representation;
- thorough;
- decisive, final and binding;
- third parties may be involved (arbitration is only between the parties);
- legal aid is available (not in arbitration or forms of ADR);
- ideal if dispute is over a point of law;
- pace of procedure can be forced through RSC Order 14/14A (no provision to force the pace of arbitration);
- strict liability may be held.

Potential Pitfalls of Litigation

- inappropriate, insufficient or inaccurate detail;
- poor presentation of case;
- parties 'lose control' to their legal representatives;
- lack of competent legal representation;
- time involved could effect decision to continue action (in the case of cashflow, *inter alia*);

Critical Success Factors in Litigation

There are few critical success factors; some believe the courts to be a 'lottery' others describe litigation as the 'legal laboratory of the courts'. Since the parties have the dispute taken out of their hands, the following critical success factors could contribute to a successful court resolution:

- competent legal representation;
- clear presentation of case;
- accurate information to the right level of detail;
- ability to afford preliminary legal costs (and any loss of cashflow) up to (and potentially beyond) the point of award.

Other Forms of Resolution

This section examines some of the less common dispute resolution procedures. Many of these constitute ADR and have been introduced to the UK from the United States. ADR is the new 'fad' of dispute resolution supported by many, including Al Gore's US National Performance Review. However it is questionable how 'new' most of the techniques are; they are simply pragmatic alternatives to some of the aforementioned procedures. The following pages describe some of the better-known methods:

Expert evaluation

This is often used as a precursor to ADR. An expert is commissioned (by both parties in agreement) to investigate the dispute and report on it. The parties may agree to settle on the report's recommendations, but sometimes it simply fuels the dispute further. The success of this form of dispute resolution will generally rest on the individuals concerned in the dispute, their attitudes and the relationship between the parties.

The Expert's decision

Strictly speaking this is not a form of ADR, although it has been present in the standard forms of contract for several years. The process is designed to pre-empt legal disputes by calling on the professional expertise of a neutral to preside on the disputed issue and form a contractually binding judgement. The process is effective in quashing the dispute, but has been considered to deny parties access to natural justice. Furthermore, pronouncing a quick and binding judgement, without full consideration of the merits, could be seen as adversarial and thus create difficulties in business relationships rather than resolve them.

Mini-trial

This is also called an 'executive tribunal' although it is not really a trial at all. The parties agree to present their dispute before a panel (usually comprising their own senior management, or a neutral panel) which makes a swift decision and issues a report of the details. Arbitration and/or Litigation may still proceed after a mini-trial, but the report is usually not admissible as evidence in any subsequent proceeding. The advantages are speed, inexpense and informality; the disadvantages include the lack of natural justice and potentially vested interests.

Tailored Arbitration

This is also known as 'med-arb', where some [usually unresolved] issues in the course of a mediation are arbitrated upon. It is rarely used, as it tends to bring out the negatives of both mediation and arbitration in a non-complementary manner: the mediator's neutrality can be lost as he is forced into the role of an adjudicator.

Evaluative mediation

As already noted some consider all mediation is evaluative to a degree. Under evaluation mediation the mediator gives a recommendation if and when so asked by the parties. It differs from 'med-arb' in that it is only a recommendation to settle. Mediators generally do not like to use this form of dispute resolution as it has the potential to undermine their neutrality.

Accelerated arbitration

Also known as 'fast track arbitration', this is built around the view that adjudication will not work if the parties are determined to resist it as it is difficult to enforce. However, it also accepts the view that arbitration has become slow, cumbersome and expensive. Accelerated arbitration is essentially an arbitration agreement with specific time duration built in. The debate surrounding this method is whether or not the award should be final and binding; especially since certain investigations will have been forced and/or hurried. The merits of accelerated arbitration are thus very similar to those of adjudication.

Instant single issue arbitration

This form of ADR is more of a concept than a practical procedure. As its name suggests it provides for a short form of arbitration which concentrates on one particular issue, rather than global 'rolled up' disputes where causation of the effects is difficult to determine. The parties are required to select the single issue in dispute, which is then taken through the accelerated arbitration route.

Documents-only arbitration

As the name suggests this is an arbitration provision based on an investigation of the documentation only; i.e. no oral evidence is considered and no cross-examination occurs. This tends to simplify and expedite the process considerably. Parties are given tight time schedules to prepare their cases, and the process is then managed by the arbitrator. The risks involved for the parties depend on the arbitrator's ability to interpret the submitted information correctly and, moreover, that the other party has not spent months preparing its case documents secretly only to spring a 'surprise dispute' on the other less-prepared party. It is suggested that this form of dispute resolution is not appropriate for larger disputes (more than £50,000 say) or contracts where little documentation is produced, such as an all-in design-build contract.

Dispute review boards (DRB)

This could be described as an 'on-site' dispute resolution panel, usually comprising three members (one nominated by each party and a third neutral nominated by the other two members). The DRB is formed pre-contract and 'retained' in the course of the works. DRB members are sent copies of major documentation and make occasional site visits to keep abreast of events. When a dispute occurs, the DRB can mobilise quickly and effectively to review the issues as a tribunal. Their decision is given as a non-binding recommendation, which in the event of dissatisfaction could be taken to arbitration.

This method has been successfully used on major contracts such as the Channel Tunnel and Hong Kong Airport and, generally speaking, it is less suitable for smaller projects. The World Bank operates a similar system employing just one member as a Disputes Review Expert. The major advantages and disadvantages lie with the speed in which the DRB can mobilise and issue its recommendation: the process is very quick, but can only be achieved through the expense of retaining the DRB 'just in case'.

Summary

In reviewing the various procedures available to resolve a construction dispute, it is clear that some procedures will be more advantageous than others depending on the circumstances of the dispute and the desired outcome.

In early 1997, the authors conducted a postal survey of the top-100 UK construction clients and the top-100 UK contracting firms to determine their preferred practices for resolving disputes. Recipients of the questionnaire were asked what features they wanted from a dispute resolution procedure (see Figure 13.3 below) and then which existing procedures they most favoured and most avoided, with reasons.

Speed	67%
Commercial pragmatism	38%
Fairness and equity	36%
A cost-effective/economic process	22%
Confidentiality	17%
Protection of business relationships	8%

Figure 13.3: What do Firms Want from a Dispute Resolution Procedure?

There were no noticeable differences between the client and contractor responses: both parties indicated that the speed of the resolution procedure is the essential requisite. This would seem to fly in the face of recent legal criticisms towards adjudication being 'rough justice'; construction professionals want quick resolutions to their disputes. Surprisingly 'cost' was of less importance, even though many see it as a concomitant of the duration. For publicly accountable organisations, the 'correctness' of the outcome was also seen to be an important issue.

When asked which dispute resolution procedure was most favoured, 81% of the respondents stated negotiation. The reasons offered were for least expense, least confrontation and maintenance of the business relationship. Conversely, when asked which dispute resolution procedure was most avoided 40% stated litigation and a further 26% opted for litigation and/or arbitration. Reasons against litigation were a unanimous response concerning the excessive costs; there was also a large proprtion of respondents who considered the process took too much time. When it is considered that, in these processes, time and expense are a corrollary of each other, the results are fairly consistent: firms want quick and cheap procedures.

However before taking this empirical evidence as a mandate from business, it is important to question whether these objectives (speed and cost-minimisation) are truly appropriate for the firm. We believe that a dispute resolution procedure will only be appropriate when it is most likely to deliver the business objectives.

Thus if the objectives are for the lowest cost of transaction and/or the quickest transaction, then these preferences will hold fast. However, where the business objectives are demanding that an alternative end-goal is achieved, then the preferred procedures must be selected with these in mind.

It is therefore necessary to consider the business objectives of the firm both overall and in the context of the transaction and disputed issue (as discussed in Chapter 2). In general the drivers on the resolution process will be at least one of the following criteria: cost and/or time minimisation (as has already been considered), sustainenance of the business relationship and/or protection of public reputation. Public bodies may have other drivers in terms of probity to consider too. Figure 13.4 reviews the most popular forms of dispute resolution against these business criteria.

	Negotiation	Mediation	Conciliation
Speed to establish	Instant: process is already in progress whether recognised or not.	Generally a few weeks.	Between a few weeks and 2 or 3 months.
Speed to resolve	Potentially very quick - at pace to suit both parties.	Quick. Usually between 2 days and 1 month.	Quick. Usually 28 days.
Expense	Negligible.	Relatively low.	Relatively low.
Confidentiality	Complete.	Complete.	Complete.
Effect on business relations	Success rests on relationship. Can be restorative.	Can be restorative.	Reduces adversary if settled amicably.
	Adjudication	Arbitration	Litigation
Speed to establish	Depends. A few weeks or months.	Can take months.	Can take years.
Speed to resolve	28 days.	Prolonged over months.	Relatively quick (days or weeks).
Expense	Generally low, if excluding retainer.	High.	High – can exceed final award.
Confidentiality	Complete.	Complete.	Public domain.
Effect on business relations	Can increase adversary.	Adversarial.	Adversarial.

Figure 13.4: Comparison of the Popular Dispute Resolution Methods.

It should be noted that often a combination of procedures is required to effect a successful resolution. Often these are pre-agreed in the conditions of contract (i.e. *ante post*), but they can be applied extra-contractually too (i.e. *ex post*) providing that they do not conflict with the express terms. Either way, it is necessary to review the full range of procedures available to choose the most appropriate procedure for the given circumstances of the firm's objectives, the business relationship, the specific contract and the dispute in issue.

One way of conceptualising the range of procedures is to view it as a 'continuum' between the extremes of negotiation (which is quick, inexpensive, and informal) and litigation (which is prolonged, expensive and formal). One extreme [negotiation] could be seen as co-operative (but *not* collaborative), where the parties control the proceedings and reach their own settlement through agreement. The other extreme [litigation] is non-co-operative and adversarial, where parties have a judgement enforced on them which is beyond their control.

The exact positions of the procedures on the continuum is contentious, but their relative positions can be plotted as shown in Figure 13.5.

The dispute is not confined to one singular position on the continuum, it migrates from left to right along the continuum, until the disputed issue is resolved. It need not include every process of dispute resolution but different procedures will be more appropriate than others at any one time.

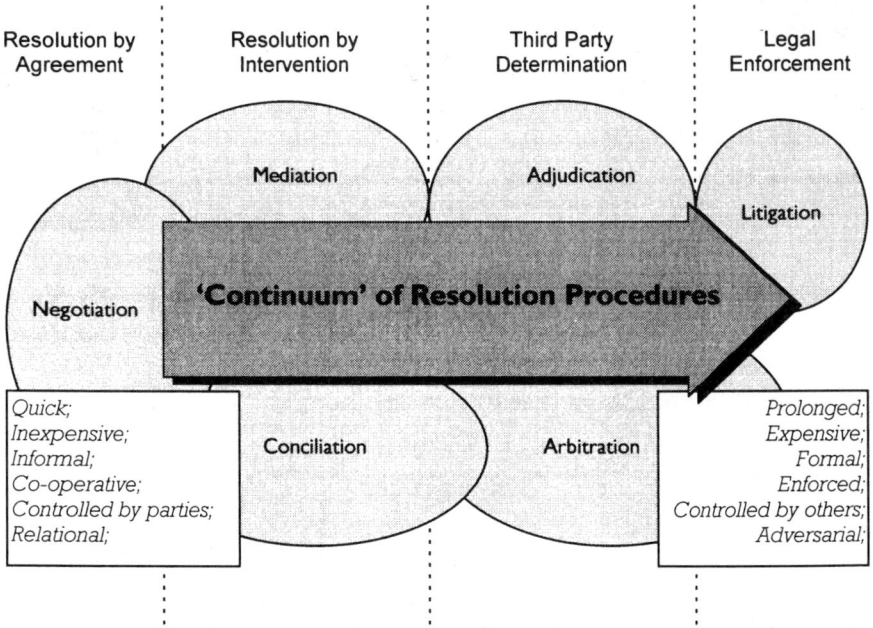

Figure 13.5: A Migration Path for Unresolved Disputes.

Given the recent rise in popularity of ADR, questions must be asked regarding whether the current resolution provisions are wholly suitable. As noted, a dispute will not pass through each of the aforementioned processes in its path to resolution; as certain steps will be inappropriate and/or mutually exclusive. For example, it has been suggested that conciliation is not compatible with mediation or adjudication and neither is arbitration with litigation. It has also been suggested that all resolution starts with negotiation whether perceived or not. There exists, therefore, a limited set of permutations for dispute resolution (leaving aside some of the other forms of ADR for the moment) as shown in Figure 13.6.

Resolution Paths	Resolution by Agreement	Resolution by Intervention	Third Party Determination	Legal Enforcement
Permutation 1	*NEG*	*MED*	*ADJ*	*LIT*
Permutation 2	*NEG*	*MED*	*ARB*	–
Permutation 3	*NEG*	*CON*	*ARB*	–
Permutation 4	*NEG*	*CON*	–	*LIT*
Permutation 5	*NEG*	–	*ADJ*	*LIT*
Permutation 6	*NEG*	–	*ARB*	–
Permutation 7	*NEG*	–	–	*LIT*

Figure 14.6: Principal Dispute Resolution Options
(Key: *NEG* = Negotiation; *MED* = Mediation; *CON* = Conciliation;
ADJ = Adjudication; *ARB* = Arbitration; *LIT* = Litigation.)

The choice of route will depend on the business drivers behind the construction work, the manner in which the approach is conducted and the relationship between the parties. For example, a dispute between parties of an arms-length contractual relationship may not have the capacity to dwell at the negotiating stages for too long, whereas a dispute within a close-working relationship is unlikely to extend beyond negotiation. However, it can be argued that it is unlikely that a generic conceptual model could be established for all governing circumstances, given the contingent nature of disputes and the differing approaches to each of the resolution processes that exist. Such a model would need to consider the over-arching business objectives of the firm, the relative degrees of power between the contracting parties, the strength of the disputed issue and both the existing and future aspired relationship between the parties.

Notwithstanding this reservation, it is possible to derive a conceptual approach to the propagation or avoidance of the dispute by considering the inherent nature of conflict in contractual relationships. The following chapter initially considers dispute avoidance and proceeds to guide the reader to a model of optimum conflict management that will serve the business objectives of the firm.

Chapter Notes

1. Some useful texts on the art of negotiation include: Fisher R., Ury W. & B. Patton (1992) *Getting to Yes, 2nd Edition*, Century Business Books, London; Steele P., Murphy J. & R. Russill (1989) *It's a Deal: A practical negotiation handbook*, McGraw-Hill Book Company, Maidenhead; and Schoenfield M.K. & R.M. Schoenfield (1991) *The McGraw-Hill 36-Hour Negotiating Course,* McGraw-Hill Inc, New York.
2. Refer to: FIDIC (1992) *Amicable Settlement of Construction Disputes*, FIDIC, Lausanne.
3. The British Academy of Experts, Committee on the Language of ADR (1992).
4. *Channel Tunnel Group* v. *Balfour Beatty* [1993] A.C. 334.
5. Definition taken from Master of the Rolls, Lord Donaldson, in an address to the London Common Law and Commercial Bar Association (1991).
6. *Cape Durasteel Ltd. v. Rosser and Russell Building Services Ltd* [1995] 46 Con LR 75.
7. Mustill M.J. & S. Boyd (1989) *Commercial Arbitration (2nd Edition)*, Butterworths, London.
8. It should be noted that US research on 220 adjudication cases has revealed that not one ajudicated decision was subsequently challenged by arbitration or litigation, as reported in the *New Civil Engineer* (14 November 1996).
9. A view put forward by Phillipp Capper (author of the NEC's Adjudication Contract) in Capper P.N. (1995) 'The Adjudicator under NEC 2nd Edition: a new approach to disputes' *Engineering Construction and Architectural Management*, Vol. 2, No. 4 (December) pp. 317 - 326.
10. Shilston A.W. (1994) 'Arbitration Revisionism' *Arbitration*, Vol. 60, No. 3.
11. Sims J. (1994) 'Incoming Chairman's Inaugral Address' *Arbitration*, Vol. 60, No. 4.
12. Herman A.H. (1994) 'The New English Draft Arbitration Bill' *Arbitration*, Vol. 60, No. 4.

13. Section 108(3) of the Housing Grants Construction & Regeneration Act 1996 requires the parties' adjudication contract to make the adjudicator's decision binding on the parties *unless* pursued by legal proceedings, arbitration or by alternative agreement.
14. Bingham T. (1994) 'Arbitral tribunals and the Courts: standards, training and supervision' *Arbitration,* Vol. 60, No. 3.
15. Murdoch J. & W. Hughes (1996) *Construction Contracts: Law and Management, 2nd Edition,* E & FN Spon, London.

Chapter 14

Conflict Management & Dispute Avoidance

Introduction

The previous chapter concentrated on methods of resolving disputes once they have occurred, i.e. a *reactive* management approach. This chapter takes a different view of conflict. It considers that conflict is a by-product of commercial relations which needs to be managed in the most appropriate way. Firstly it looks at ways in which disputes can be avoided (i.e. a *proactive* management approach). Many texts on contract administration and dispute resolution suggest that dispute avoidance is the most desirable situation and that approaches which minimise conflict will directly contribute to achieving the overall performance targets of the firm. This view was strongly supported by the human relations movement in the 1960s and 1970s and implied that conflict should be avoided and/or eliminated.

Although this is an admirable proposition, it does not necessarily follow that dispute avoidance directly leads to enhanced performance and, furthermore, there is little empirically-verified data to support such statements. Indeed, many industrial psychologists have written on the benefits which conflict can bring (such as enhanced performance, motivation, creativity, change, control and/or group cohesion). One commentator has even observed that: *"conflict, effectively managed, is a necessary precondition for creativity"*[1].

This is not to suggest that conflict should be promoted under all circumstances. Chapter 12 demonstrated the incredible expense of the current levels of conflict in the construction industry. Clearly there is a need to establish an optimum balance between conflict propagation and conflict avoidance. Thus, this chapter proceeds to consider how conflict can be managed *effectively* in order to raise performance levels and maximise the overall business objectives.

Dispute Avoidance

As presented in Chapter 12, conflict can be described in an escalation process from the establishment of an initial problem through to its 'manifestation' as a dispute (see Figure 12.1). *Dispute Avoidance* comprises the process(es) required to prevent conflict manifesting as a dispute; it is also referred to as *Dispute Prevention*[2].

Chapter 12 identified two contrary schools of thought regarding conflict. The first, that conflict and disputes should be resolved as efficiently as possible, holds to the philosophy of dispute avoidance. There are two approaches to dispute avoidance that are commonly promoted:

1. The **good project management** approach which suggests that disputes can be avoided through better management of the construction process.
2. The **project partnering** approach which suggests that closer-working relationships are the solutions to dispute avoidance.

In the following, these two approaches are considered in detail to determine whether this 'efficiency' approach is valid as a business proposition.

Good project management

Good project management is at the heart of successful construction management. Construction is a project-based industry and without competent project management skills, the industry would lose control of the way in which works are procured and delivered to the client.

Project management skills are multi-various and will include many of the following: supplier assessment and accreditation, contract negotiation and administration, preparation of quality-controlled specifications, resource allocation and control, supervision of contractors' activities, programming and progress monitoring, change control, financial management and supply co-ordination. Each one of these skills is aimed at enhancing the way the project delivery is controlled; in essence project management is a *control* mechanism on the construction process.

That control and good project management is critical to the successful delivery of constructed works is not at question. However the focus of conflict management that suggests disputes can be avoided by managing the construction process in a better way needs to be considered a little more closely.

There is an implication in this way of thinking that suggests that disputes only occur as a result of errors and/or non-conformities in the project documentation and contract management. As such the parties begin to disagree on an interpretation of the documentation and the process of conflict escalation begins. It therefore follows, in this argument, that the irregularities and discrepancies are the cause of the conflict and that better project management would eradicate these errors, therefore eliminating conflict and increasing the likelihood of achieving the business objectives. Thus, to follow this line of argument, good project management avoids conflict and this leads to greater business performance in construction.

In essence, this approach belongs to the same school of thought as Total Quality Management and the New Engineering Contract (NEC). Both methods offer management solutions which are aimed at improving the overall delivery process, i.e. by 'getting it right first time'. That these methods will, on occasions, contribute to the reduction of claimsmanship and conflict is undeniable, but whether they will go as far as to prevent all conflict is highly questionable. While good project management clearly offers advantages in reducing error and/or discrepancy, it still fails to address the root causes of dispute. In general, error and discrepancy are used as 'vehicles' of conflict; the root cause usually comprises other *behavioural* aspects that result from parties seeking their own self-interests. Consequently, any approach that suggests conflict is avoidable through good project management denies basic business fundamentals (concerning the realisation of an individual's own commercial interests). While good project management will be contributory to the avoidance of disputes, it will not address the root causes of conflict and thus will not prevent disputes from occurring on all occasions.

Project partnering

The other popular approach proposes that ***project partnering*** and closer-working business relationships are the solutions to dispute avoidance. This view, which was elaborated on in Chapter 4, has been endorsed by Sir Michael Latham and other industry commentators[3] (see, for example, Case Study 10 overleaf). It supposes that if clients were to adopt collaborative relationships with their suppliers then there would be no disputes and everyone would achieve their business objectives. This view belongs to the 'win-win' school of thought and is akin to the 'lean supply' concept which has been argued to apply to construction activities[4].

While it is true that disputes may be reduced and/or even alleviated through project partnering, the approach fails to consider the overarching business objectives of the parties to the transaction.

The focus has slipped away from performance and, in its place, the elimination of conflict has become one of the primary goals of the firm. Thus, in other words, the concept has confused *means* and *end*. In general, by attempting to find 'win-win' solutions to overcome conflict, a firm will be compromising on its business objectives. It assumes that there is equality of action and purpose within the supply chain and that, primarily, it is in every firm's interests to support each other. While this may be appropriate under a certain (albeit peculiar) set of circumstances, it does not generally follow for either all or the majority of cases. The argument presented by advocates of project partnering suggests that, instead of aggressively pursuing its business objectives, a firm should be prepared to compromise on its goals in order to maintain the *Pax Romana*.

Such philosophy, while being admirably charitable, does not serve the interests of the individual, unless, as previously noted, specific conditions apply. Furthermore, the vehicle of conflict elimination (purported predominantly to be that of *trust*) does not account for the self-interest and power struggles over scarce resources which prevail in all business relations (elsewhere referred to as 'the human condition'). The partnering concept fails to account for opportunistic attitudes prevalent in industry and the risk of the 'free-rider' within the supply chain. Project partnering may alleviate conflict between parties, but this will be at the price of achieving business success[5].

Case Study 10: Partnering in the United States

Much talk has been given to the contribution of partnering in construction. The evidence of its success and the debates over its definitions have, however, been major stumbling blocks to its acceptance. The advocates of partnering have been quick to look to the US for empirical justification of this method. While the following three vignettes point to partnering's success, it should also be noted that these examples are not representative of all types of construction. The following clients are able to offer regular workload to their preferred suppliers and thus create an on-going sustainable business relationship. This link between business relationship and contracting methods is explored in greater detail in the final section of this book.

1. The *US Army Corps of Engineers* has been operating partnering agreements since 1988. In the first six years of implementation (i.e. in the course of 200 projects) contractual claims had reduced by more than 60%, none of which has gone to litigation.

2. Since 1991, the *Arizona Department of Transportation* has been partnering projects worth upto $300,000 in value. Of the first 100 projects procured this way, there were no claims whatsoever and reported cost savings of $5 million in total.

3. The *Los Angeles County Metropolitan Transportation Authority* began to employ partnering on a series of subway project phases and saw an immediate drop in claims of 10%.

Source: Ellison & Miller[6]

Despite industry-wide recognition of these two approaches to dispute prevention (project management and project partnering), neither can be accepted as being a commercially valid panacea of conflict in their totality, for the reasons already stated. To develop a robust concept of conflict management, we first need to consider the business objectives of the firm, the balance of power and the structural conditions of the market which exist as part of the background to the commercial transaction(s) on which the elements of conflict are founded. From here, it is possible to develop a theoretical framework for effective conflict management in the context of the business relationship that pre-exists.

Effective Conflict Management

The early chapters of this text considered that the *raison d'être* of the firm is the creation of a profit margin within its particular market sector and that this has been overlooked by many when considering supply chain relationships. In order to achieve business success, construction activities (whether core or support) should be aligned to the strategic intent of the firm in such a way as to *maximise* this end-goal. Therefore the issue is not whether conflict exists, or whether it can be minimised, it is about how conflict can be *optimised* to serve the overall business intent; i.e. how it can be managed in the most 'effective' manner.

Thus conflict *minimisation* should not become an end-state, a point which has already been laboured. There are many commentators who consider conflict carries both constructive and destructive qualities. The key to conflict management is to understand how conflict contributes to achieving the business goals and thus how it is to be managed. If the use of conflict serves the business case and delivers the required benefits then it must be adopted. Conversely, if it adds cost and wastes valuable resources, which consequently detract from achieving the business goals, then it must be avoided.

Conflict (and its propagation or avoidance) thus becomes a supply chain management issue that is subject to the over-determination of sourcing strategies. This includes the nature of the business relationship and a consideration of the relative degrees of [idiosyncratic] power between the parties to the contract[7].

It follows that relationships that are arms-length and outwardly 'adversarial' by nature will encourage conflict; indeed a **conflict of interests** between the parties is inherent in this type of business relations. Thus transactions established from competitive price-based tendering and other multiple-sourcing strategies are implicitly adopting an adversarial approach to commerce. The buyer will attempt to lever prices down, while the supplier will attempt to counter-avail prices upwards. Conflict is established at the very root of the transaction. The fallacy of the law of contract is to suppose that offer and acceptance replaces commercial conflict with an 'agreement'.

This view can be supported by inter-organisational studies, where it has been observed that: *"Without a common goal individuals are, in effect, licensed to do the best for themselves within the rules"*[8]. Similarly, one construction commentary correctly asserts that: *"...conflict occurs when* [the contracting parties'] *objectives are incompatible"*[9].

Thus there is a correlation between the 'community' of the parties' goals and the degree to which conflict is present. As Charles Handy has observed: *"When two or more groups interact with differing goals, sets of priorities or standards, there is likely to be conflict. The higher the degree of interdependence of the units, the more crucial becomes the relationship of their objectives and ideologies."*[10].

This leads the discussion to consider how conflict is effected by the converse, i.e. when business relations are close-working. These type of relations entertain a **coincidence of interests** between the parties and thus remove the likelihood of conflict being manifested (the parties are jointly working to 'mutual goals'). That is not say that conflict does not exist. It is recognised that even in the most hierarchical governance structures conflict is still present[11]. However when a 'coincidence of interests' exists, parties will jointly seek to minimise and/or avoid conflict manifesting as it is in their shared interests to do so; as such conflict is absolved.

Thus there exists a framework for conflict management. Where the association of interests between parties is one of conflict, conflict will be manifested. Dispute avoidance measures will only serve to reduce the appropriation of the business objectives of the commercial exchange. Conflict should be expected and when it is experienced, it needs to be resolved in such a way that maximises the individual's own interests.

However, conversely, where there is a co-incidence of interests, there exists a proactive framework of conflict absolution. Disputes are prevented *ex ante*, the parties are collaborating to such an end that serves their mutual objectives and they jointly seek to avoid disputes.

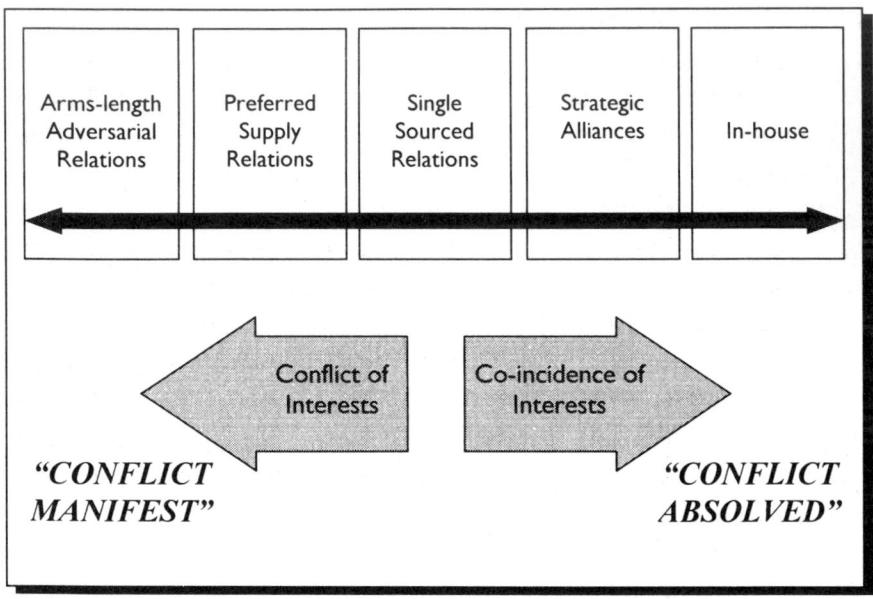

Figure 14.1: Conflict and the Association of Interests.

This can be illustrated conceptually by over-laying this framework for conflict management on an established model of business relations (see Figure 14.1). This figure shows a continuum of contractual relations and the link these relations have with conflict management. This continuum of relations determines the interim forms of contractual governance between 'market' structures (multiple sourced arms-length relations) and the 'hierarchy' (in-house relations). It is based on the *relational competence* approach[12] to controlling assets and resources of varying degrees of specificity to the organisation, based on the seminal works of Coase, Williamson and Reve[13].

Thus it is the selection of a contractual relationship, that is fit-for-the-purpose of achieving an individual firm's business goals, which will determine whether disputes are avoided or not. In arguing that different relationships will be fit-for-purpose, it can also be argued that conflict propagation will, at times, be fit-for-purpose and, thus, the management of conflict becomes a concomitant of managing the business relationship.

Empirical Verification

In order to verify/falsify this conceptual framework, a considerable degree of empirical research needs to be conducted. It would need to establish whether there are causal links between the governing relationship, the degree of conflict (manifest in disagreements, claims and/or disputes) and whether the firm's business objectives have been achieved. A considerable amount of data has already been presented in the course of this text. This includes the literature search from business and construction management texts, a synthesis of the construction industry's existing disputes management practices and a review of published research data to support this text. However none of this data is suitable for testing whether a causal link exists between the business relationship and level of conflict, as suggested.

In Chapter 13, some of the results of a national survey of construction disputes were presented. The following section of this chapter presents some of the remaining data from that survey in support of the verification/falsification of the fore-going model of conflict management. The purpose of the survey was to establish contemporary levels of dispute and adversarialism in the construction industry and to identify current dispute management practices. In particular it considered the efficacy of the existing resolution procedures and how disputes can be proactively avoided (if at all). Not all of the results are relevant to this chapter and, therefore, the following interpretation of the survey data has concentrated on the *relational* aspects of the parties' transaction and their impact, if any, on the level of explicit conflict. It should be noted that this evidence, like other empirical studies, is only as accurate as the degree to which the data was representative of the respondents' corporate practices at the time of the survey.

The survey was conducted in February 1997 and the responses represented a little over 20% of the construction industry's market (based on Department of the Environment statistics). It was expected that all survey recipients, as regular employers/ contractors of construction, would have experience of contractual claims and disputes with other parties and that, furthermore, few companies could claim to be 'dispute-free'. However 9% of the respondents claimed that they were 'never involved in disputes' thus potentially undermining the presumption that conflict exists in all business transactions. The reasons suggested for not having disputes could be summarised under two main headings of: those who elected to contract in a 'non-adversarial' manner and those who adopted 'partnering' practices. Unfortunately this does little to test the fore-going concept, as it could not be determined whether these practices were compromising on, or helping maximise, the broader business objectives. The only deduction that can be drawn is that, in practice, the application of non-adversarial and/or collaborative practices helps avoid conflict being manifest, as previously suggested.

One of the biggest problems with the use of the word 'partnering' is its definition, as discussed in Chapter 4. A significant proportion of respondents (upto 85%) considered partnering to be one of the greatest opportunities of reducing adversarialism and dispute in the industry. However no two respondents could offer the same definition for the term, which has significant consequences. If two parties to a contract implicitly believe they are 'partnering' and yet are unable to explicitly define what is meant by the term, there is immediate potential for divergent intent within the contract. Conflict is more likely to exist; it is only the parties' implicit belief-system on what constitutes partnering that prevents conflict manifesting.

The constituent parts of each respondent's definition indicated some common themes, but also that a diverse understanding prevails. The references to 'working together', 'team working' and 'repeat business' shows a partial collective towards collaboration; but this is insufficient to suggest that this comprises one monolithic relationship 'type'. It would not be possible to place this on our continuum of 'fit-for-purpose' contractual relations, except to suggest that these relationships exist somewhere between the extremes of arms-length contracting and in-house supply. That the industry believes these relationships will offer 'the greatest opportunity' to reduce dispute and absolve conflict would appear to support (albeit weakly) the former concept (Figure 14.1), only inasmuch as conflict is manifest under the 'traditional' arms-length approach to construction procurement. To make any further suppositions would constitute gross manipulation of this evidence.

From the survey results, it was clear that the 'people factor' had an important bearing on matters of conflict. A third of respondents considered that the 'wrong' people administering a contract was one of the *main root causes* of dispute. This represented the single-most common root cause of all responses. But what does this mean? Again it is not possible to deduce what respondents mean by 'wrong' or what the effect on the business goals might be. (For instance, does wrong refer to those who will adversarially pursue a disputed issue, or to those who prefer to collude in order to avoid conflict? Under certain circumstances either option might be appropriate or, conversely, equally inappropriate!) Nevertheless the implication in this result is that there is a perceived 'human' *relational* factor which effects the causation of disputes.

Furthermore, 88% respondents considered that disputes are a product of the way the construction industry conducts its business relations. Most (61%) considered that open tendering has created an adversarial culture which gives rise to conflict and that it is the people involved, rather than the contracts, who create disputes. Finally, when asked whether it was any specific form of contract which promoted adversarialism, 67% respondents preferred to state that none did in itself and that it was the people who administer the contract who promote adversarialism; thus reinforcing the *relational* impact on conflict causation.

When questioned on success criteria for dispute resolution methods, the over-riding desire was for quick procedures (69%) as reported in Chapter 13. Of note, was a small percentage of respondents who considered that 'protection of the business relationship' was an important criterion for dispute resolution.

This was further supported by respondents' first-preference of negotiation as being the 'least confrontational' dispute resolution technique, having the ability to 'maintain business relationships' with the other party. Furthermore some respondents said they preferred to avoid litigation because of the 'bad feeling' it creates between parties.

Clearly the business relationship is an important factor not only in resolving disputes, but also in managing conflict generally. The 'means' in which disputes are resolved are perceived to have an effect on the on-going business relationship. Linked to the former evidence, there is now a clearer understanding that the business relationship *does* affect the level of conflict between parties, but that also the way in which the conflict is handled has a consequential effect on the business relationship.

This point was reinforced in the survey results, by suggesting the commercial relationship has a tangible effect on a dispute if there is the likelihood of future business. Only 5% of clients said that a dispute would *not* effect their decision to do business with a contractor again; while only 7% of contractors would not change their view on an existing dispute if there was the incentive of future workload.

It is evident that given the opportunity to take a longer-term view of an on-going workload, both client and contractor will view an existing dispute differently. Although not wholly conclusive, this evidence provides strong suggestion that clients and contractors can establish a *coincidence of interests* through incentive-based contracting and thus absolve conflict. It suggests that there is a strong *relational* influence on the management of conflict.

Summary and Conclusions

Therefore to summarise this chapter, it has been established that conflict can be managed in the context of an over-determining business relationship between the parties to a transaction. It has been argued that dispute avoidance is not necessarily preferable; instead conflict should be managed effectively. Further, it can be argued that there is a place for conflict when it suits the business objectives. The manifestation of conflict is a direct product of the underlying interests within a transaction. This association of parties' interests, when opposing, will produce conflict and, when co-incidental, will absolve conflict. Thus when there is a governing close-working relationship that creates and sustains a *coincidence of interests*, disputes can be avoided.

This conclusion forms an integral part of the praxis offered to construction professionals in this publication. By analysing the emotive subject of disputes in the construction industry, it has been shown that effective management of an over-determining business relationship can govern the performance outcomes of the transaction.

Chapter Notes

1. Hall J. (1971) Decisions, decisions, decisions *Psychology Today* (November) pp. 5, 51-54, 86-88.
2. Construction Industry Institute (1995) *Dispute prevention and Resolution techniques in the Construction Industry*, CII, The University of Texas at Austin.
3. Refer to: Latham M. (1994) *Constructing the Team*, HMSO, London; Bennett J. & S. Jayes (1995) *Trusting the Team*, University of Reading; Baden-Hellard R. (1995) *Project Partnering: Principles and Practice*, Thomas Telford, London; Construction Industry Institute (1995) *Dispute Prevention and Resolution Techniques in the Construction Industry*, CII, The University of Texas at Austin; Ellison S.D. & D.W. Miller (1995) 'Beyond ADR: Working Toward Synergistic Strategic Partnership' *Journal of Management in Engineering*, Vol. 11, No. 6.
4. Refer to: Womack J.P. & D.T. Jones (1996) *Lean Thinking*, Simon & Schuster, New York, p. 292.
5. For more discussion on the fallacies of lean thinking refer to: Cox A. (1997) *Business Success: A Way of Thinking About Strategy, Critical Supply Chain Assets and Operational Best Practice*, Earlsgate Press, Boston, UK.
6. Ellison S.D. & D.W. Miller (1995) 'Beyond ADR: Working Toward Synergistic Strategic Partnership' *Journal of Management in Engineering*, Vol. 11, No. 6, pp. 44 - 54.
7. Refer to Cox, *Op. cit.* and also: Watson G. & J. Sanderson (1997) 'Collective Good versus Private Interest' in Cox A. & P. Hines (eds.) *Advanced Supply Management: The Best Practice Debate*, Earlsgate Press, Boston, UK.
8. Handy C. (1993) *Understanding Organisations, 4th Edition*, Penguin Books, London, p. 309.

9. Murdoch J. & W. Hughes (1996) *Construction Contracts: Law and Management, 2nd Edition*, E & FN Spon, London, p. 367.
10. Handy, *Op. cit*, p. 300.
11. See for example: Fox A. (1966) *Industrial Sociology and Industrial Relations*, HMSO, London; Lawrence P.R. & J.W. Lorsch (1969) *Developing Organisations: Diagnosis and Action*, Addison-Wesley, New York; Pascale R.T. (1990) *Managing on the Edge: How Successful Companies use Conflict to Stay Ahead*, Penguin Books, London.
12. Refer to: Cox A. (1996) 'Relational Competence and Strategic Procurement Management: Towards an Entrepreneurial and Contractual Theory of the Firm' *European Journal of Purchasing & Supply Management*, Vol. 2, No. 1, pp. 57 -70.
13. Coase R. (1937) 'The Nature of the Firm' *Economica*, Vol. 4, pp. 386 - 405; Williamson O.E. (1975) *Markets and Hierarchies: Analysis and Antitrust Implications*, Free Press, New York; Reve T. (1990) 'The Firm as an Internal and External Nexus of Contracts' in Aoki M., *et al.* (eds.) *The Firm as a Nexus of Treaties*, Sage, London.

Part E

Contracting for Business Success

Chapter 15

Fit-for-Purpose Contractual Relations

Introduction

Preceding sections of this book have examined both the background to contracting in the construction industry and the various ways in which it can be carried out. The purpose has been to equip the reader with knowledge and understanding about the specific contractual details required to procure construction works effectively. Chapter 1 referred to this as *contracting competence*. In the chapters that followed this, each of the commonly-used standard forms of contract have been examined. It is assumed that this analysis provided an understanding of the basic principles of each contractual mechanism. To do this two types of analysis were conducted:

1. *Allocation of risk* – each contract apportions responsibilities and liability to the parties to a contract for actions and events that occur during the progress of the works. From this, it follows that there are both 'winners' and 'losers' depending on which events occur, what resultant remedies are prescribed and how parties in the contract are affected.

2. **Structure of power** – each contract specifies a control mechanism to which all parties are tied-in contractually. Within this structure parties are given different powers (rights and responsibilities) to act and make decisions. These have a consequential effect on the way in which the contract is administered and the allocation of value within it. Again, there are 'winners' and 'losers' depending how the controls are structured and how they are administered during the course of the works.

The preceding chapters have outlined the results of this approach for each type of individual contract. The purpose of this final chapter is to draw together the results of this analysis into a single overview. By so doing, it is intended that the reader will be provided with a framework within which 'fit for purpose' contractual relations can be determined.

The concept of 'fit for purpose' implies that there are specific commercial objectives which are required by the purchaser when procuring construction works. These are usually seen in terms of cost, quality, functionality, time to market, etc. It follows that these objectives should be reflected in the out-turning works contract and business relationship between buyer and supplier and, moreover, that the contract and relations are structured in such a way as to ensure these business objectives can be fulfilled in the most optimum manner.

What should be perfectly clear to the reader by now, however, is that the current range of construction contracts have unique properties in the ways in which they are structured and, thus, they will be suited to fulfilling different objectives. Only certain forms of contract will be appropriate for use under some specific conditions, according to the goals of the works that are to be procured.

This chapter begins to address these differences and seeks to establish a framework within which the construction manager may select a contract which is 'fit for purpose'. In particular it looks at each of the contracts that have been analysed previously to draw a comparison with their structural properties. This overview then proceeds to consider the effect that different generic clauses have on a commercial relationship and whether the current forms of construction contract drive the parties to close-working collaborative relations or, alternatively, towards arms-length non-collaborative relations. From this, a conceptual framework of 'fit for purpose' contractual relations can be derived.

It is important to stress that, while providing the reader with a conceptual understanding of contracting for business success, this chapter does not attempt to prescribe a model which tells the reader how to select the most appropriate contract. This specific need has been covered in detail in the parallel text by the authors. This other book (*The Contract Selection Toolkit*[1]) provides a simple step-by-step methodology for construction professionals to choose, from the array of published standard forms, a contract that most suits his/her particular needs. This methodology has been designed to suit a number of different applications across a wide variety of business environments, as well as supplying important practical guidance on each of the issues involved. We recommend that practitioners seeking hands-on guidance to select the most appropriate form of construction contract should make first-hand reference to this parallel text (further details are to be found at the end of this chapter).

Risk Apportionment

Within the construction industry there are two popular views of risk apportionment that are regularly espoused. The first suggests that contracts should be drafted with an even balance of risk in their clauses, while the second suggests that risk should be allocated to the party most able to manage it. From this current thinking, notions of 'fairness' and 'equity' are considered as being better practice for contract management. In the following, it is argued that this thinking is only partially correct.

The notion of contractual fairness was first supported by Sir Michael Latham in his review of UK contracting and procurement arrangements in construction. Since then, the Chartered Institute of Purchasing & Supply (CIPS) has published a 'best practice charter' with the Construction Round Table (CRT) called *Commitment to Fair Construction Contracts*. Clients and contractors have been advised to sign up to it and agree to abide by its principles (see below). The hope is that, by this, the industry will be rid of pernicious practices, such as pay-when-paid clauses.

Signatories to the Commitment to Fair Construction Contracts have given notice that they intend to conform with the following principles:

- To deal fairly with those with whom we contract and with those who are contracted to them in an atmosphere of mutual co-operation.
- To work together (recognising that we and the parties with whom we contract are members of a team) and to seek and adopt 'win-win' solutions to problems which arise.
- To move toward the use of an integrated and compatible family of contracts which:
    - is suitable for the full ranges of types of construction contracts and procurement routes;
    - is written in language which is easy to understand; and
    - clearly defines the roles and duties of everybody who contributes to the project.
- To allocate risks between the parties on a rational basis so that each risk is allocated to the party best able assess, carry and manage it.
- To minimise changes to the planned works. Wherever possible, the impact of each change on the conduct of work will be forecast and any consequent changes in the price agreed before the work is done, or if not before, as soon as possible afterwards.
- Whenever possible, to base the interim payments on agreed schedules of payments, milestones or activities including payments for off-site activity.
- To provide that the period for making interim payments (which is to be stated in the contract) shall be as short as administratively possible.
- To make interim payments in the period in which they become due as agreed in our contracts and to provide compensation for delay in making any payment by adding interest at an appropriate rate.
- Whenever possible, to provide rewards for exceptionally good performance in our contracts, which should be defined at the start of the work.
- To minimise the risk of disputes arising in the management of the construction work and provide competent, third party resolution of disputes which we cannot settle ourselves.
- We look to those with whom we contract to discuss any case where the full adoption of these principles has not been made and we shall:
    - consider each case; and
    - where justified, take appropriate action.
- To commit all our staff who are involved in construction projects to adopting these principles. We call upon those organisations with whom we contract to adopt the same principles with those with whom they contract in connection with all construction projects. In this way, we expect that the principles will come to be applied throughout the pyramid of organisations contributing to all construction work.

Figure 15.1: Commitment to Fair Construction Contracts.

Source: CIPS/CRT

But do such initiatives actually work? The Fair Contracts charter may have successfully raised the profile of a handful of large construction firms, but what long term effects has it produced? The evidence, two years on, appears to be very little. This charter received only luke-warm support from a few well-meaning members of the industry and there has been no evidence to suggest any change in behaviour or performance has resulted from its implementation. The conclusion that has to be drawn from this initiative is that it was merely a token gesture that failed to reach the root causes of business behaviour in commercial transactions in the industry.

One key question that remains unaddressed in these contemporary calls for fairness is whether this is an appropriate approach for improving the industry's performance. While some degree of regulation is undoubtedly required, these appeals for fairness are, in the view of the authors, misguided. Furthermore, they demonstrate a fundamental misunderstanding of basic commercial principles.

A mandate for fairness is an appealing concept on the surface but, at heart, it works against the principles of entrepreneurship and commerce. Such a mandate assumes that parties are agreeing to 'unfair' terms and conditions in a contract against their own wishes. Yet, it can be argued that such notions run contrary to the weight of case history developed in contract law over hundreds of years. Moreover, in attempting to mandate a fair balance of risk, these calls are denying practitioners the opportunity to make more profitable returns by taking measured risks in their endeavours. This illustrates one of the founding principles of business: a party undertakes to carry the risk of a certain activity and, upon completion, is duly reimbursed as a reward.

It can be argued therefore, that, prior to any pre-contractual negotiations, buyers of goods or services should establish, on the basis of their own competence and entrepreneurial flair, which risks they are able to accept and which risks they want to pass on to the supplier. This then becomes the subject of pre-contractual negotiations in which the supplier factors in a premium to his/her prices in order to accept the risk that is to be managed. This will be priced on the basis of the supplier's competence and entrepreneurial ability. If agreement cannot be reached then it is likely that the supplier does not have the necessary degree of competence to accept the risk and supply the buyer's requirements.

Thus a commercial precedent has been set throughout history and has been, when necessary, regulated by the Courts. The recent calls for 'fairness', however, suggest that these principles are not working and that risk apportionment in contracts should be more 'balanced'. If the risk is balanced, then it is suggested that the playing field is levelled too. However, is this appropriate? What happens if a buyer or supplier does not want to accept the balance that has been mandated to it? (Or, equally, what if it is unable to accept those terms?). What happens if either party can get a better deal from an unbalanced apportionment of risk? In other words, just because we wish to be thought of as acting fairly (on the basis of someone else's views), should we deny the right to procure (or supply) construction works more effectively? Should commercial transactions be restricted in this way if both parties understand the extent of their contractual liabilities?

These are thought-provoking questions and there are no quick or easy answers. It is the authors' view that the recent attempts to make contracts 'fair', within the broader context of improving industry performance, may be fundamentally misguided, however well-meaning and appealing the message. It can be argued that the industry ought, first, to re-consider the ways in which construction is procured and how this can be made to be more effective, before it considers to seek improvements in construction performance through notions of equity.

Risk allocation in the Contracts

In each of the contracts studied in preceding chapters, the specific allocation of risk has been analysed and described. Figure 15.2 (overleaf) illustrates a summary of this work by considering how much risk each party carries when construction works are procured using a specific form of contract. The diagram considers, in general terms, the whole of the procurement process and not just the impact of the main contract. The effect of this allocation on business relations and the way in which they are conducted is discussed later in this chapter. But for now, Figure 15.2 demonstrates the fact that most of the current standard forms of contract are relatively 'balanced' (despite what other commentators might say) between client and contractor. In general, clients accept the risk of their works specification and the known condition of their site, while contractors accepts the risk of their operations and workmanship. Moreover, no party in the process is devoid of all risk, however good their internal risk management and contract negotiation skills.

	Clients	Professional Advisers	Contractors
Sequential Contracting Method			
ICE 6th			
FIDIC 4th			
JCT'80			
Design & Build Contracting Method			
ICE D-C			
IChemE Red			
JCT'81			
Minor Works Contracting Methods			
ICE MW			
JCT IFC'84			
Other Contracting Methods			
NEC			
JCT MC/87			
IChemE Green			
GC/Works/1			
DEFCON 2000			
Construction Management			

Key:

Low Risk   High Risk

Figure 15.2: A Summary of Risk Allocation.

It should be noted that, for most types of construction procurement, the professional advisers (architects, engineers, quantity surveyors, etc.) carry very little risk. Although the doctrine of privity of contract applies for the express terms of the construction contract, Figure 15.2 illustrates general risk apportionment for the whole procurement method and not just the specific form of contract used. Thus, it can be argued that the professional bodies carry the least risk of all parties for their participation in the construction process. This is an important point which will be picked up later in the text.

Structures of Power

The other major analysis throughout this study of contracts has been the structures of power that are built into the commercial transaction. This structure is made manifest in two principal ways. Firstly there is a control mechanism within each contract which assigns rights and responsibilities to each party to make decisions that have a material effect on the performance of the contract. Examples of this include responsibilities under the contract to award interim payments, to make changes, to add/omit work items or to terminate the contract. This list is not exhaustive and simply provides an insight to what is meant by the term *control mechanism*. The second form of power structure is an over-determining mechanism that effects the behaviour and actions of a party. Examples of this include the use of incentive structures and/or penalty clauses. Later in this chapter, it is shown that the business relationship has a strong over-determining influence on the way in which the contract is discharged.

It is through the control mechanism that the client should be ensuring that the works are being delivered to time, to cost and to the right specification of quality. An inadequate power structure in the contract will not ensure that these criteria are met.

Herein lies a particularly complex issue regarding the type and extent of controls required for the delivery of the works. In each of the forms of contract examined in this book, these controls have been documented and discussed. However it is impossible to specify, in advance, the precise manner of control that is required.

This will depend on the competence of the contractor, the type of commercial relationship that exists and the nature of the transaction that is being undertaken.

It is for this reason that the standard forms of contract assign rights to individuals within their terms to control the way in which the works are constructed. The named individuals then have the responsibility to take up their powers, or not (as appropriate) during the course of the works.

However, as indicated in the preceding chapters, the construction industry has developed a habit of continuing to empower third party professional agents to control the performance of the contract. This principle of tri-lateral governance was discussed in detail in Chapter 5. While this may be seen as a pragmatic solution to administering the contract, it is flawed for a number reasons:

1. Independence from the contract can rarely be assumed. Although legally set apart from the discharge of contractual duties by the doctrine of privity of contract, the third party is usually party to a separate commercial agreement with one of the parties to the main contract (the client). In these circumstances independence is difficult to assume, however 'professional' the adviser may be.

2. Impartiality is nearly always impossible to achieve for the third party, mainly due to the fact that independence cannot be assumed automatically (as above) and also that many of the contracts specifically express decisions to be *"...in the opinion of the [third party]"* thus encouraging subjective views to be formed. Many of the professional institutions contend that their members are professionals and thus are able to stand apart of such subjectivity but, in practice, this is difficult to accept when everyone knows that the professional is on the payroll of one of the parties to the main contract.

3. The parties to the main contract are losing control of the administration of the contract and subjecting themselves to the decisions of external bodies. Not only does this create additional transaction costs, it actually means that the parties experience a loss of control. The problem comes when the

client wants to take a certain course of action and is subsequently told by the third party agent that, under the terms of the contract, they are unable to do so.

Examples of these points can be seen in most of the standard forms of contract, such as the roles of the architect in the JCT'80, the engineer in the ICE 6th Edition and the project manager in the NEC and/or GC/Works/1 contracts.

	Clients	Professional Advisers	Contractors
Sequential Contracting Method			
ICE 6th			
FIDIC 4th			
JCT'80			
Design & Build Contracting Method			
ICE D-C			
IChemE Red			
JCT'81			
Minor Works Contracting Methods			
ICE MW			
JCT IFC'84			
Other Contracting Methods			
NEC			
JCT MC/87			
IChemE Green			
GC/Works/1			
DEFCON 2000			
Construction Management			

Key:

Little Power   Much Power

Figure 15.3: A Summary of Contract Power Structures.

Indeed, a synthesis of the previous analyses of each form of contract would suggest that the third party professional is placed in a particularly favourable position of power by the way in which these contracts are structured. Figure 15.3 illustrates this point clearly by providing a summary cross-reference of the allocation of power in each of the standard forms of contract. The general results are unequivocal: virtually all contracts assign the majority of power to the third party professional bodies.

Ironically, the majority of forms of contract are drafted by professional institutions which have promoted the role of their own membership in every contract. This is successfully implemented by giving each third party considerable authority to act, without the corresponding risk or liability for their actions (as noted earlier in Figure 15.2). The result may be a form of protectionism maintained by the industry's professional institutions in the interests of their own members and not necessarily in the interests of the rest of the industry.

Let us consider this in closer detail with the example of the Institution of Civil Engineers (ICE). The ICE publishes contracts which require the presence of a third party engineer to supervise the administration of the contract. In the ICE contracts, considerable power and authority is vested in the Engineer (see page 98, for example); however, where are the liabilities or risks carried by the Engineer? To date, the ICE has not published a contract for the engagement of an Engineer when using its own forms of contract. Thus, it would appear that the ICE promotes the power and authority of its membership without making them directly accountable for their actions in any of their written contracts. It successfully does this by taking the decision-making from the parties to the contract (particularly the client) and assigning powers to a third party engineer. In short, one might conclude that these contracts appear to assign power without accountability.

In defence of the ICE, its contracts are drafted by an industry committee which includes, apart from itself, the Association of Consulting Engineers and the Federation of Civil Engineering Contractors (which recently disbanded). The ICE would maintain that this removes subjective interest from their forms of contract by giving an industry-wide representation and, thus, a 'balanced' structure. This, however, remains contested: two groups of professional engineers and the former-support of a contractors' group hardly represents the full diversity of the industry. Whether the ICE contracts are 'fair' and 'balanced' probably rests on the individual views of the parties to the contract at any given time during commercial proceedings.

Finally in this sub-section, a thought must be given to the way in which specific clauses or groups of clauses in any form of contract influence behaviour and performance. Obvious examples, as noted earlier, are the use of penalty clauses and/or incentives. These are used, to greater or lesser extents, in over-determining ways to shape the actions of parties to the contract. These types of structures have limited success rates because of their effect on the business relationship, as will be shown later. For example, in a collaborative relationship it may be more appropriate to incentivise good performance with rewards (e.g. a bonus for early completion), rather than disincentivise poor performance with penalties (e.g. liquidated damages for delay). As will be demonstrated in the following section of this chapter, it is important to use the appropriate conditions of contract with the appropriate business relationship to gain maximum benefit of both mechanisms. Like the principles of super-imposition, an inappropriate match will have detrimental effects on both the relationship and the contract performance.

Contractual Relations

In Chapter 2, the notion of *contractual relations* was introduced. This term links two complementary concepts of external resource management: the conditions of contract (which is predominantly what this book has been about) and the business relationship between buyer and supplier.

To date, very little academic work has been done in the construction management field to understand the 'softer' [less tangible] principles of an over-determining business relationship. Most established research and development work has concentrated on the 'harder' [more tangible] principles of contract administration, project management, costing models and risk management. More recently, work has begun to seek to understand the benefits of 'partnering' but this appears to have stalled with disagreements over the concept's definition and application.

Interestingly enough, the field of purchasing and supply management has approached the concept of contractual relations from the other end. Since the mid-1980s, there has been a proliferation of literature espousing the benefits of relationship management. The evidence to support its benefits has, to date, been anecdotal and somewhat inconclusive. Again, effective reasoning and supportive evidence (in factual and counter-factual terms) has been inhibited due to semantic squabbles and tactical marketing campaigns.

Despite the lack of clear indisputable evidence, there are indications that there are benefits of a relational approach to contracting in specific circumstances on certain occasions. The term *contractual relations* attempts not just to bridge the two ideas but, more fundamentally, to establish that there are key associations to be established between the contract and the relationship.

In Chapter 2 it was noted that the business-specific qualities of an exchange relationship have an determining effect on the contractual matters of the transaction (page 49). But, what does this mean? Put simply, a pre-existing business relationship between Client and Contractor may have a considerable influence on the way in which construction works are procured. For example, a supplier who knows that a regular workload could be secured just by getting each contract built to cost and on-time, will be motivated to ensure this is the case. In this example the repeat episodes of each individual contract combine to form an over-determining relationship. The parties to this particular relationship may prefer to put aside certain specific contractual rights or obligations within an individual contract in order to preserve the broader business relationship. One example might be the waiver of rights to interest on an overdue payment in the interests of future work. In this way, the relationship that has developed between buyer and supplier has over-determined the specific conditions of contract.

This is not say that all types of business relationship have beneficial or favourable effects on the contractual performance.

Consider the effects of aggressive non-collaborative arms-length relations when asking (say) to waive certain pecuniary rights or to work in trust and co-operation while assessing the potential costs of additional works.

It is clear, just from these notional examples, that contractual relations need to be considered carefully to ensure the right qualities are brought together to procure works effectively. There are four principal tenets which comprise contractual relations and, as shown in Figure 15.4, these comprise: the relationship itself and the three main components of the contract (the risk allocation, the responsibilities and the reimbursement mechanism).

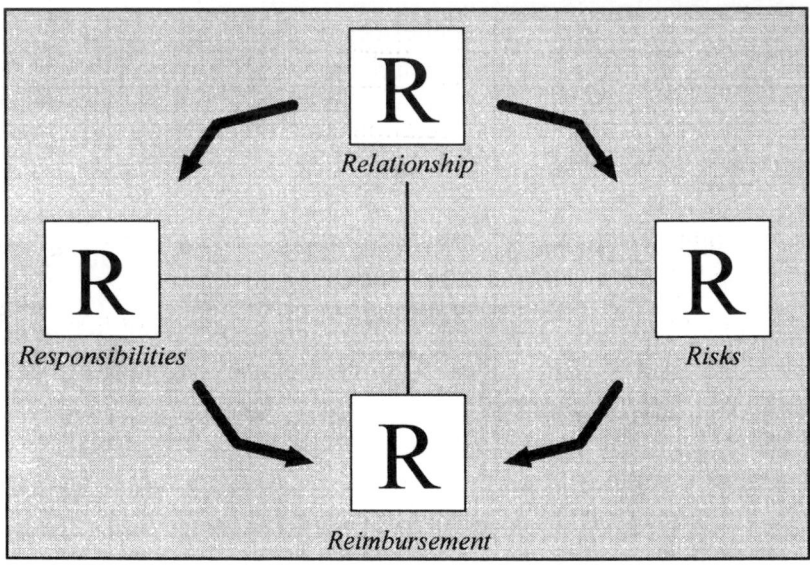

Figure 15.4: The 4 R's of Contractual Relations.

These have been codified simply as the 4 R's of contractual relations. In short, the relationship is seen to have an over-determining effect on the way in which the transaction is conducted. As noted in the aforementioned example, the commercial relationship influences the responsibilities (rights) of each party in accordance with the inherent power structure which that relationship has. It also effects the degree of risk sharing (liabilities) which parties may be willing to accept. Instead of becoming a contested issue throughout pre- and post-contractual negotiations, the allocation of risk will change in response to the type of relationship that is effectively in place. For example, a collaborative relationship is likely to enjoy a higher degree of risk sharing. In the same vein, arms-length relations tend towards strict liability and passing of risk.

It follows from these effects, that there will be a corresponding effect on the payment terms (reimbursement) agreed between buyer and supplier. Thus the 4 R's: the relationship determines the responsibilities and the risks, which in turn determine the reimbursement.

Types of Relationship

There are several fundamental questions that remain outstanding from the above discussion. These include what is it that drives the relationship, how is the relationship determined and, once identified, how can the conditions of contract be selected from this commercial relationship?

Throughout the text in this book, the authors have clearly expressed the need to establish relationships that are 'fit-for-purpose'. Fitness for purpose has a specific legal definition and it is from this meaning that the term has been adopted in commercial relations. If a commodity has been identified as having a specific function that has been clearly expressed and articulated, then that commodity is said to be fit-for-purpose when it clearly conforms to fulfilling that defined function (purpose). Similarly business relationships will be fit-for-purpose when their original function is fulfilled.

But, what is the purpose of a business relationship? Any casual observer of contemporary construction management literature could be mistaken for thinking that the purpose of relationships is to (1) avoid conflict with other parties; (2) develop trust by being more co-operative with others; (3) to work together as a team; and/or (4) be more considerate and accommodating to others. In short the driver has come from the notion of 'win-win' – i.e. that, we would all get along a lot better if we could have equal shares of the cake.

Of course, in commercial terms, this is nonsense! Firms do not exist primarily to share the dividend of their operations with other companies; they exist to make profitable returns for themselves *only*. Collaborative working can only be advocated when it can be seen demonstrably to be a more effective means of appropriating a profitable return for oneself! Thus the purpose of commercial relationships is to ensure that the business objectives of the transaction(s) can be met in a controlled way.

This issue of control is an important concept and one that varies with the range of available business relationships. Figure 15.5 indicates a continuum of buyer-supplier relationships and the range of control which they can offer. The diagram is based on the concept of *relational competence*[2]. It indicates the link between the degree of control a certain business relationship can offer and the type of contractual form that is required.

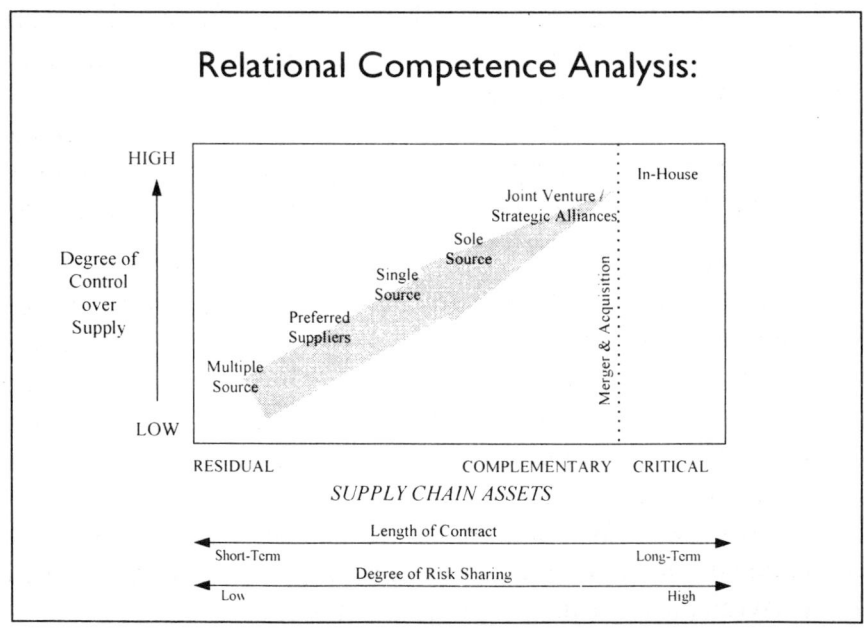

Figure 15.5: A Range of 'Fit-for-Purpose' Relationships.

Source: adapted from Cox[2]

Multiple sourced relations have tended to prevail in construction. These are often called arms-length relationships as the client treats each transaction in a discrete once-off manner and competitively tenders each one from a wide source of suppliers. Contracts tend to be short-term (i.e. for the duration of the project only) and there is little risk sharing (strict liability prevails). It is contended that through these types of transaction, the purchaser is given very little control on the works and it is mostly left to a reactive and tactical approach of writing tight specifications and 'policing' contractual compliance.

Conversely, highly collaborative relations, found in strategic alliances and joint ventures (whether equity-sharing or not), carry a considerable degree of shared risk and are, usually, conducted over a long time horizon. The purchaser is able to have greater control, without the need of excessive on-costs to police compliance, as the supplier is working towards the same objective and a 'coincidence of interests' is prevalent (as noted in Chapter 14).

Another way of defining these close-working relations is to suggest that if they were to become any 'closer' or more collaborative, then they would become comparatively indistinct from in-house commercial relations.

The remaining types of relationship in the relational competence model comprise the range of external relations between multiple sourced and strategic alliance. In short, these are:

- **Preferred Supply** – these relations occur when the client elects to restrict the vendor-base to a small number of 'preferred suppliers' who are able to offer products and services to the client on preferential terms. These types of activities and resources are complementary to the clients business and competition is limited while collaboration is encouraged in order to enhance the performance of the supply offering. Control over the supply-base is greater than that of open competition, as the preferred suppliers strive to maintain their preferential status and a regular workload.

- **Single Source** – these relations have many similarities to the above, except that the client elects to restrict the vendor-base to just one single supplier. While the products or services may be highly complementary to both the buyer's business operations, there are obvious dangers of developing an over-dependence on this chosen supplier. This then leads to a loss of control on costs, quality and innovation.

- *Sole Source* – these relations are the same as single sourced relations, except that the vendor-base is not limited to one sole supplier by choice because there is only one supplier in this particular market.

It should be noted that this continuum of relationships deliberately avoids the use of the 'partnering' and 'partnership sourcing' terms. It is suggested that these terms simply confuse construction professionals by the diverse definitions they have been given. Using the model in Figure 15.5, these approaches could be anything from preferred supply relations through to shared-equity joint ventures. The problem is that, without industry-wide recognition and acceptance, these terms are relatively meaningless. A more acceptable approach would be to define relations by the relative degrees of collaboration and 'closeness' of relations between the buyer and supplier, as in the approach taken by relational competence.

The Existing Forms of Contract

In the course of the research that has led to the publication of this book, the authors conducted a postal survey of regular construction clients in the UK. In total, 163 clients were surveyed which represented the views and practices of approximately 16% of the national construction market. Clients were asked when they used different contracts, how they made their selection, whether the contracts were amended and what were clients' views of the contracts[3].

Although space precludes full dissemination of the results here, there were some interesting and potentially alarming trends that prevailed. For example, when asked about their satisfaction with the existing standard forms of contract, only 6% of respondents were found to be fully satisfied with their provisions. Furthermore, only 5% were found to be satisfied with the existing allocation of risk in the standard forms of contract, while only 23% claimed to be constantly provided with the required results by them. At the same time, a surprising 47% of clients claimed to be experiencing disputes over the wording of these contracts. For some, this has created the situation where they had stopped using particular forms.

This high level of dissatisfaction was supported by survey findings about the extent to which clients attempt to use other contractual solutions. Over 90% of respondents claimed to amend the standard contract clauses to suit their own requirements, of which 87% did it to re-allocate specific risks to the contractor. Similarly a high level of use of bespoke contracts was observed, with 47% respondents claiming to use their own drafted documents.

The two major conclusions drawn from this survey are that, firstly, construction contracting practices are very traditional and, secondly, clients are extremely dissatisfied with the available forms of contract.

To understand why these contracts appear to be giving so little satisfaction, an analysis was made of each form[4]. Putting on one side the inappropriate allocations of risk and power (as already commented), this analysis also establishes that the vast majority of construction contracts are written to support arms-length relations and contractual compliance. The general principles in these forms of contract suggest that if a contractual requirement is not fulfilled then the standard recourse is usually in the form the recovery of damages or penalties.

Typical examples where this is evident are found in the provisions of liquidated damages for delayed performance, the use of performance bonds and retention monies for inadequate performance and the use of the rights of 'set-off'. The tendency in these contracts is to push the risk attached to responsibilities into a *strict liability* situation – i.e. to ensure that the liability for a given set of actions belongs wholly to one party to the contract or wholly to the other. There are very few opportunities for risk and reward sharing schemes, very few incentives to enhance performance and little opportunity to promote collaborative working.

The conclusion from this study was that arms-length non-collaborative behaviour is being actively encouraged by most standard forms of contract. Furthermore, rather than being a mechanism to unite buyer and supplier in a common cause (i.e. to construct the works), the contract was being used as a wedge to drive distance between them (see Figure 15.6).

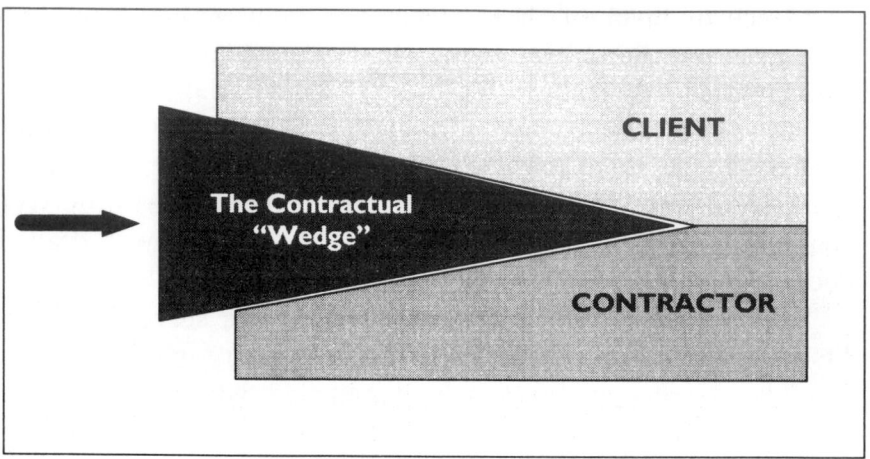

Figure 15.6: The Contract as a Wedge Driven Between Parties.

Figure 15.7 (overleaf) charts the standard forms of contract that were examined and their provisions to apportion liability and/or incentivise collaborative behaviour. Although the measures recorded here are not fully inclusive, they do nevertheless reveal some distinct trends.

Measures that support arms-length contractual relations include the use of disincentives, or *negative incentives*[5]. These are measures that penalise the contractor if the contract is not complied with. The four categories used in Figure 15.7 illustrate ways in which the client is able to exert pressure on the contractor to comply with the express requirements of the contract. The use of the Performance Bond and retention monies operate in similar ways: if the contractor fails to comply, then the client has the means and the right to take or withhold money from him. Similarly, damages are 'awarded' to the client if the contractor defaults in his responsibilities; the client then has the right to deduct money from the contractor's payment either directly or, on occasions, by set-off against other payments. These four measures treat the contractor at arms-length and do little to instil collaboration.

Measures that support collaborative contractual relations tend to use incentives rather than disincentives to unite the parties in a co-incidence of interests. The four categories used in Figure 15.7 illustrate ways in which the client draws the contractor closer by providing the incentive to work collaboratively. The emphasis is on shared responsibility and shared risk or reward. This could include sharing cost-savings, co-working through problems on-site and/or principled agreements to share risks rather than seek liability.

Contract:	Measures providing support for collaboration				Measures militating against collaboration			
	Incentive schemes	Risk sharing schemes	Shared rewards	Joint working	Use of Bond	Damages or penalties	Retention of money	Set-off/ withheld payment
JCT'80						✓	✓	✓
JCT'81						✓	✓	✓
JCT IFC'84						✓	✓	
JCT MC/87						✓	✓	
FIDIC 4th	O				✓	✓	✓	
ICE 6th					✓	✓	✓	
ICE D&C					✓	✓	✓	
ICE MW						✓	✓	
NEC	O	O	O	✓	O	O	O	
IChemE Red						✓	✓	
IChemE Green				✓		✓		
GC/Works/1			✓	✓		✓		✓
DEFCON 2000			✓		O	✓	✓	✓

[KEY: O = optional clause; ✓ = expressed requirement in clauses.]

Figure 15.7: Analysis of Contract Clauses on Collaboration.

At a glance, Figure 15.7 indicates a general preference among standard forms of contract for measures that militate against collaborative relations, rather than those which support them. One conclusion to be drawn from this analysis is that the majority of construction contracts are *reactive* mechanisms designed to apportion blame as and when non-compliance occurs. Moreover, collaborative relations are unlikely to be fostered if clients continue to use these types of contract.

A potential solution would, therefore, appear to be the introduction of positive incentives to all standard forms of contract. However, previous research on incentives in construction contracts in the United States has revealed that although the general performance is marginally greater (by perception), there are significantly more disputes and contractual disagreements associated with these conditions[5]. Clients tended to use incentives as an inducement for contractors to accept greater risk. The research empirically shows that the inclusion of incentive clauses simply exacerbates the adversarial nature of the industry; this time with increased claims concerning whether or not the improved performance target was achieved and the incentive rightfully gained.

Thus, the general conclusions from this analysis are that:

- the majority of standard forms of contract are *reactive* mechanisms designed to apportion blame between the parties;
- collaborative relations are unlikely to be fostered if clients continue to use contracts based on disincentives and blame-apportionment;
- collaborative relations are unlikely to be fostered even if incentives are included in contracts
- a balance, therefore, needs to be established between the type of relationship and the contracting strategy that is adopted.

These conclusions support earlier statements related to the correlation between the contract and the relationship. If parties continue to focus on the contract alone without due regard to an appropriate over-determining relationship, then arms-length adversarial behaviour (and all its effects) will prevail; a balance between the contract and relationship needs to be established.

In the previous sub-section of this chapter, a typology of contractual relations was presented. In a similar vein, a corresponding typology of contracts can be developed which typically comprise:

- *conditions of contract* designed to apportion risk for every possible circumstance;
- a governing *head of agreement* supported with standard form of contract;
- negotiated terms and conditions governed by a *head of agreement*;
- a [negotiated] *memorandum of understanding* with no express terms or conditions;
- an *oral contract*.

The various methods of contracting business that exist could be attributed to each of these five categories. For example commodity 'spot' purchases are usually contracted through a purchase order of standard terms and conditions. Similarly, competitive open tendering for contracts is usually supported with a full set of express terms and conditions to ensure the risks associated with every eventuality are covered. In these two examples the first category *conditions of contract* apply and business is usually associated with arms-length behaviour. It is in this category that most forms of construction contracts are found.

The second two categories of contract imply a reduced dependence on the written conditions of contract and the introduction of relational factors associated with preferred supply sourcing. The *head of agreement*, sometimes referred to as a framework agreement, can be contractual (as in a term agreement or call-off contract) or non-contractual (as in a teaming agreement). When the importance of the relationship between the parties is greater and there is more focus on collaborative working, jointly agreed [negotiated] contract terms will be more appropriate.

In the situation where suppliers are highly integrated in the form of an extended enterprise and high degrees of reciprocity and interdependence exist, there is little need for a contract despite the distinction between organisations. This is the virtual company structure of vertical quasi-integration. The parties do not need express terms and conditions to govern their responsibilities as *"...the issues of ownership, control and power become incredibly blurred and confused"*. Instead, a joint understanding is negotiated and a 'coincidence of interest' prevails: the contract need be nothing more than a *Memorandum of Understanding* and carries little legal force.

In the case of strategic alliances, joint ventures and shared equity ownership, a contract is not always necessary. Distinction needs to be made between the contract required to form the alliance and the individual transaction episodes for the works. The individual transactions do not require expressly defined contracts as ownership is shared: an *oral contract* will suffice.

Conceptual Framework

From these types of contract and types of relationship, a conceptual framework of 'fit-for-purpose' contractual relations can be developed. This is best described in the following illustration (Figure 15.8, overleaf). Clearly there are linking factors between the relationship and the type of contract and a direct correlation must not be assumed. These factors might include the relative competence of the supply-base, the frequency of transactions, the complexity of the works content and/or the degree of opportunism present in either party. Although space precludes a detailed analysis of these linking factors, our parallel text (*The Contract Selection Toolkit*) provides a step-by-step methodology which helps practitioners select the appropriate type of contract for the prevailing circumstances.

This model can be developed further to 'map' the correlation between the type of contract and the type of relationship. As already noted the correlation is not going to be direct; there are other intervening variables which effect the appropriate link between the business relationship and the type of contract that is used.

Nevertheless, empirical evidence from the contracting practices of 70 UK construction clients has suggested that there is a correlation between the type of contract and the type of relationship[4]. That is, as the business relationship becomes more collaborative there is less of a need for formal conditions of contract. Indeed, it is already widely disseminated that some clients, such as McDonalds Restaurants and Rover, do not use standard forms of contract to procure their works requirements. Instead they prefer to adopt purchase order agreements (mainly for accounts payable purposes) and control the progress of the works through the structural conditions of an over-determining relationship. Both these clients are examined in detail in a parallel text by Cox and Townsend[6].

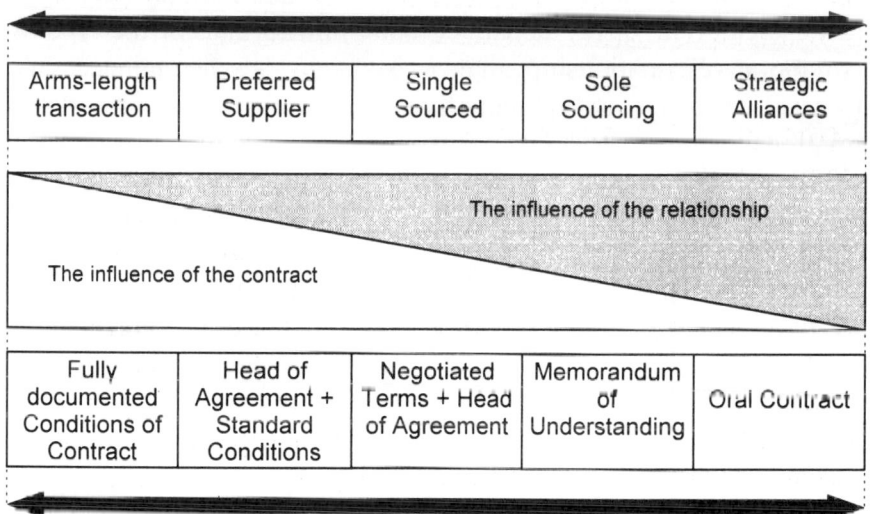

Figure 15.8: A Conceptual Model for Contractual Relations.

In general terms, the evidence from the industry supports the presence of a causal link between business relationship and the contract type. Figure 15.9 (overleaf) presents this link by illustrating a theoretical 'ideal' of best practice. There are, as with any models of best practice, practical restraints which make this ideal difficult to implement in practice. Contracting strategies tend to be far more complex than this simple classification and correlation indicate. It is not always possible to differentiate between the contract and its application from the relationship; contractual relations lack distinct resolution and should be considered as holistic concepts and not necessarily disaggregated in this way.

That is not to say that a distinction cannot be drawn: clearly it is possible to misalign relational and contractual practices (as in the example of arms-length non-collaborative relations with the use of the NEC). Furthermore, if misalignment does occur, there will be clear ramifications in terms of an effect on the works performance. Figure 15.9 suggests that this can occur in one of two ways: either (1) a contract that is too informal for the type of non-collaborative relationship may place the parties in a position of unacceptable commercial risk, or (2) a contract that is too formal for the type of collaborative relationship that exists may create unnecessary transaction costs and potentially jeopardise the close-working relations.

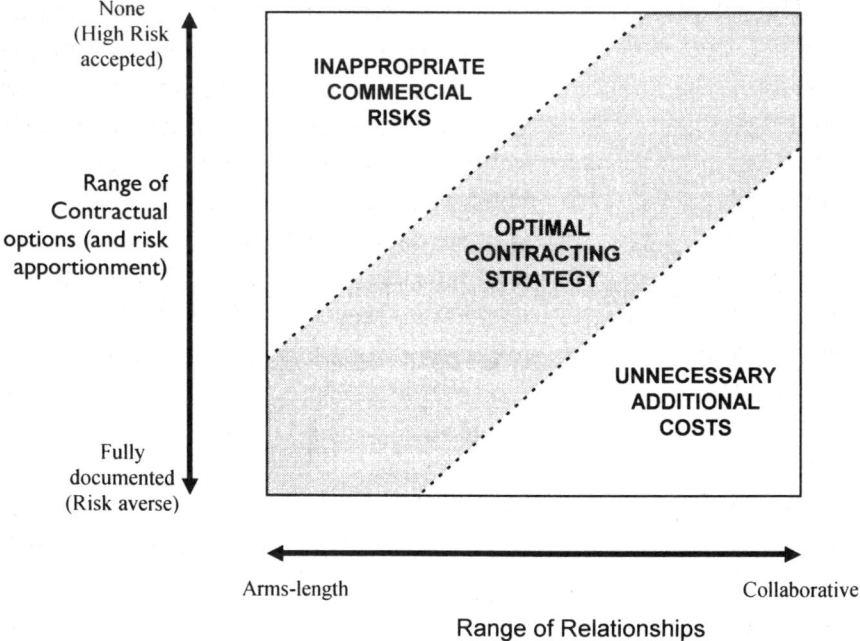

Figure 15.9: The risks of sub-optimal contracting strategies.

Contract Selection

Given this correlation between the choice of business relationship and the type of contract and the subsequent effect on the performance of the works, there is clear unequivocal need for construction clients to select the most appropriate form of contract with care and precision. In this context, a construction client refers to any organisation within the construction supply chain that is purchasing goods or services from the industry and not just the end-user clients such as Railtrack, BAA or Whitbread.

This chapter has begun to draft out a conceptual framework for the establishment of 'fit-for-purpose' contractual relations. It has clearly demonstrated that clients should not consider the conditions of contract in isolation from other factors. Clearly there is a need to determine an appropriate business relationship with the supply-base prior to selecting whichever form of contract is considered necessary for the works.

The selection of the form of contract (if it is a form of contract that is required) will be determined by several interrelated variables. In Figure 15.4, some of these variables have been prescribed as the *responsibilities* (and powers) to act within the terms of the contract, the *risk allocation* between the parties and the *reimbursement* structure for the value exchange.

However there are other variables that practitioners need to account for, as well as these factors. The authors' research has demonstrated that, in practice, the choice of contract is determined by a complex thought-process which may include measures such as: the technical complexity of the works, the choice of contracting strategy, the degrees of uncertainty that exist and the amount of control that is required.

Practical guidance is limited in the industry and few decision-making tools exist (hence the need for the research which led to the publication of this book and its parallel texts). Those that have been published are limited to broad and inconclusive generalisations that remain open to the subjective interpretations (and abuse) of the user.

Concluding Statement

This chapter has summarised the preceding chapters by taking a broad, but critical, analysis of the provisions within the standard forms of construction contract. In many places this analysis has demonstrated significant inadequacies in these contracts as commercial instruments.

If the interpretations from this analysis are to be accepted, it follows that it is not surprising that so few practitioners achieve effective construction procurement.

Many of the standard forms of contract are written by professional institutions to *endorse* and *protect* the role of their own members in the procurement process. The 'endorsement' occurs by giving the contract a structure of power that encompasses the need for a supervising third party professional who is then given tremendous powers of authority to act at the expense of the contracted parties. The 'protection' in the contracts occurs through the doctrine of privity of contract; few of the professional bodies are made liable for their actions within the terms of the contract. The results are clear:

> *Most construction contracts allocate power and responsibility to those with no corresponding accountability or liability for their actions.*

This misappropriation of power is most clearly shown when considering the effect on the parties to the contract. In agreeing to be bound by the express terms of the contract, most clients and contractors are surrendering control over the administration of the contract while still accepting all the normal commercial risks associated with construction works.

The problems of contracting in the construction industry are further exacerbated by some of the current practices that are promoted. It is clear that little consideration is given to the potentially detrimental effect of the conditions of contract on general business relations and overall performance. However, it is equally evident that, little consideration is given to the over-determining effects of the business relationship on the way in which the contract is administered and the works are procured effectively.

This concluding chapter has sought to demonstrate that further thought, throughout the industry, should be given to the interaction of contractual relations and their effect on performance. Earlier this was described as **contracting competence** where *understanding* is required in order to apply the most appropriate contracting solution to the particular circumstances of the transaction. For this, *knowledge* is required both of the full range of tools and techniques that are available to practitioners and of the total universe of circumstances in which they could be applied, as well as strategic clarity of the goals that an organisation wishes to achieve.

It is hoped that this book has provided the reader with, at least, a clearer understanding of the tools and techniques that are available in contracting, as well as some understanding of their appropriateness for application.

Further details and a step-by-step methodology to select the optimum form of construction contract can be found in our parallel text: *The Contract Selection Toolkit*.

Chapter Notes

1. I. Thompson & A. Cox (1998) *The Contract Selection Toolkit: Practical Guidance for the Construction Industry*, Earlsgate Press, Boston, UK.
2. Cox A. (1996) 'Relational Competence and Strategic Procurement Management: towards an entrepreneurial theory of the firm' *European Journal of Purchasing & Supply Management*, Vol. 2, No. 1, pp. 57 - 75.
3. Cox A. & I. Thompson (1998) 'Has Contracting lost its Customer Focus?' *Journal of Construction Procurement*, Volume 4 (May).
4. The results of this analysis were first disseminated by the authors in a paper given to the 6th Annual IPSERA conference and subsequently published in: Thompson I., A. Cox & L. Anderson (1998) 'Contracting Strategies in the Project Environment' *European Journal of Purchasing & Supply Management*, Volume 4, Special Issue, (March).
5. Ashley D. B. & Workman B. W. (1986) *Incentives in Construction Contracts*, Construction Industry Institute, Austin, Texas.
6. Cox A. & M. Townsend (1998) *Strategic Procurement in Construction*, Thomas Telford, London.

Index

Learning Resources
Centre